EASY STEPS TO SUCCESS: A GRAPHING CALCULATOR GUIDE

TO ACCOMPANY

MATHEMATICAL APPLICATIONS
FOR THE MANAGEMENT, LIFE, AND SOCIAL SCIENCES
SEVENTH EDITION

HARSHBARGER/REYNOLDS

LISA S. YOCCO
Georgia Southern University

RONALD J. HARSHBARGER
University of South Carolina Beaufort

HOUGHTON MIFFLIN COMPANY   BOSTON   NEW YORK

Sponsoring Editor: Lauren Schultz
Assistant Editor: Marika Hoe
Manufacturing Manager: Florence Cadran
Marketing Manager: Danielle Potvin

Copyright © 2004 by Ronald J. Harshbarger and Lisa S. Yocco. All rights reserved.

No part of this work may be reproduced or transmitted in any form or by any means, electronic or mechanical, including photocopying and recording, or by any information storage or retrieval system without the prior written permission of Houghton Mifflin Company unless such copying is expressly permitted by federal copyright law. Address inquiries to College Permissions, Houghton Mifflin Company, 222 Berkeley Street, Boston, MA 02116-3764.

Printed in the U.S.A.

ISBN: 0-618-29364-7

1 2 3 4 5 6 7 8 9 – MA – 07 06 05 04 03

PREFACE

Easy Steps to Success: A Graphing Calculator Guide for the TI-83, TI-83 Plus, and TI-82, gives step-by-step keystrokes and instructions for these calculators, along with examples using these keystrokes and instructions to solve problems. The split screen format shows the menus and keystrokes needed to perform or to check algebraic operations on the left with step-by-step examples of the use of these menus and keystrokes on the right.

This guide is written as a calculator supplement for *Mathematical Applications for the Management, Life, and Social Sciences,* 7th Edition, by Ronald J. Harshbarger and James J. Reynolds. The *Mathematical Applications* text is designed to give flexibility in the use of technology, so this guide presents all steps for each topic presented. Because many of the keystrokes and menus are used repeatedly throughout *Mathematical Applications*, the topics in this guide are not matched to specific sections of the text, but are presented in a logical order consistent with the text. This permits easy access to the calculator keystrokes and menus as they are needed, often repeatedly, in the text. The appropriate topics can be found easily by referring to the Contents or Index of this guide.

This guide begins with an introduction of the keys and important menus of the TI-83, TI-83 Plus, and TI-82 graphing calculators, followed by the step-by-step procedures and examples for the topics of *Mathematical Applications*. These topics include arithmetic calculations, graphing equations and functions, finding intercepts of graphs, solving equations and systems of equations, evaluating algebraic expressions and functions, finding domains and ranges of functions, finding vertices of parabolas, finding maxima and minima of polynomial functions, modeling data, solving problems with matrices, solving inequalities, and solving problems involving sequences, probability, and descriptive statistics.

Many of the keystrokes and menus are identical on the TI-83, TI-83 Plus, and TI-82, so the instructions given on the pages of this guide are for the TI-83, TI-83 Plus, and the TI-82. In those cases where the instructions differ, the special instructions will be stated clearly and identified as being for the TI-83, TI-83 Plus, or the TI-82. This permits students with different calculators to work together collaboratively.

Lisa S. Yocco

Ronald J. Harshbarger

CONTENTS

OPERATING THE TI-83	1
OPERATING THE TI-83 PLUS	2
OPERATING THE TI-82	9
I. CALCULATIONS WITH THE TI-83 AND TI-82	14
Calculations	14
Calculations with Radicals and Rational Exponents	15
II. EVALUATING ALGEBRAIC EXPRESSIONS	16
Evaluating Algebraic Expressions Containing One or More Variables	16
III. GRAPHING EQUATIONS	17
Using a Graphing Calculator to Graph an Equation	17
Viewing Windows	18
Finding y-Values for Specific Values of x	19
Graphing Equations on Paper	20
Using a Graphing Calculator to Graph Equations Containing y^2	21
IV. EVALUATING FUNCTIONS	22
Evaluating Functions with TRACE, Value	22
Evaluating Functions with TABLE	23
Evaluating Functions with Y-VARS	24
Evaluating Functions of Several Variables	24
V. DOMAINS AND RANGES OF FUNCTIONS; COMBINATIONS OF FUNCTIONS	25
Finding or Verifying Domains and Ranges of Functions	25
Combinations of Functions	26
Composition of Functions	27
VI. FINDING INTERCEPTS OF GRAPHS	28
Finding or Approximating y- and x-Intercepts of a Graph Using TRACE	28
Using CALC, ZERO to Find the x-Intercepts of a Graph	29
VII. SOLVING EQUATIONS	30
Using TRACE to Find or Check Solutions of Equations	30
Solving Equations with the ZERO (or ROOT) Method	31
Solving Equations Using the Intersect Method	32
Solving An Equation in One Variable with SOLVER on the TI-83 and TI-83 Plus	33
Solving an Equation for One of Several Values with SOLVER	33
VIII. SOLVING SYSTEMS OF EQUATIONS	34
Points of Intersection of Graphs - Solving a System of Two Linear Equations Graphically	34
Solution of Systems of Equations Using the Intersect Method	35
Solution of Systems of Equations – Finding or Approximating Using TABLE	36

IX. SPECIAL FUNCTIONS; QUADRATIC FUNCTIONS	37
Graphs of Special Functions	37
Approximating the Vertex of a Parabola with TRACE	38
Finding the Vertex of Parabolas with CALC, MAXIMUM or MINIMUM	39
X. PIECEWISE-DEFINED FUNCTIONS	40
Graphing Piecewise-Defined Functions	40
XI. SCATTERPLOTS AND MODELING DATA	41
Scatterplots of Data	41
Modeling Data	42
XII. MATRICES	43
Entering Data into Matrices; The Identity Matrix	43
Operations with Matrices	44
Multiplying Two Matrices	45
Finding the Inverse of a Matrix	46
Determinant of a Matrix; Transpose of a Matrix	47
Solving Systems of Linear Equations with Unique Solutions	48
Solution of Systems of Three Linear Equations in Three Variables	49
Solution of Systems - Reduced Echelon Form on the TI-83 and TI-83 Plus	50
Solution of Systems of Equations: Non-Unique Solutions	51
XIII. SOLVING INEQUALITIES	52
Solving Linear Inequalities	52
Solving Systems of Linear Inequalities	53
Solving Quadratic Inequalities	54
Solving Quadratic Inequalities - Alternate Method	55
XIV. LINEAR PROGRAMMING	56
Graphical Solution of Linear Programming Problems	56
XV. EXPONENTIAL AND LOGARITHMIC FUNCTIONS	57
Graphs of Exponential and Logarithmic Functions	57
Inverse Functions	58
Exponential Regression	59
Alternate Forms of Exponential Functions	60
Logarithmic Regression	61
XVI. SEQUENCES	62
Evaluating a Sequence	62
Arithmetic Sequences - nth Terms and Sums	63
Geometric Sequences - nth Terms And Sums	64
XVII. MATHEMATICS OF FINANCE	65
Future Value of an Investment	65
Future Values of Annuities and Payments Into Sinking Funds	66
Present Value Formulas - Evaluating with TABLE	67
SOLVER and Finance Formulas on the TI-83 and TI-83 Plus	68
Annuities and Loans - The FINANCE Key on the TI-83	69

XVIII. COUNTING AND PROBABILITY — 70
- Permutations and Combinations — 70
- Probability Using Permutations and Combinations — 71
- Evaluating Markov Chains; Finding Steady-State Vectors — 72

XIX. STATISTICS — 73
- Histograms — 73
- Descriptive Statistics — 74
- Probability Distributions with the TI-83 and TI-83 Plus — 75

XX. LIMITS — 76
- Limits — 76
- Limits with Piecewise-Defined Functions — 77
- Limits as $x \to \infty$ — 78

XXI. NUMERICAL DERIVATIVES — 79
- Numerical Derivatives — 79
- Checking a Derivative — 80
- Finding and Testing Second Derivatives — 81

XXII. CRITICAL VALUES — 82
- Critical Values — 82

XXIII. RELATIVE MAXIMA AND RELATIVE MINIMA — 83
- Relative Maxima and Relative Minima Using The Derivative — 83
- Relative Maxima And Minima Using Maximum and Minimum — 84
- Undefined Derivatives and Relative Extrema — 85

XXIV. INDEFINITE INTEGRALS — 86
- Checking Indefinite Integrals — 86
- Families of Functions - Solving Initial Value Problems — 87

XXV. DEFINITE INTEGRALS — 88
- Approximating a Definite Integral-Areas Under Curves — 88
- Approximating a Definite Integral-Alternate Method — 88
- Area Between Two Curves — 89

INDEX — 90

Operating the TI-83

TURNING THE CALCULATOR ON AND OFF	[ON]	Turns the calculator on
	[2nd] [ON]	Turns the calculator off
ADJUSTING THE DISPLAY CONTRAST	[2nd] [▲]	Increases the display (darkens the screen)
	[2nd] [▼]	Decreases the contrast (lightens the screen)

Note: If the display begins to dim (especially during calculations), and you must adjust the contrast to 8 or 9 in order to see the screen, batteries are low and you should replace them soon.

The TI-83 keyboard is divided into four zones: graphing keys, editing keys, advanced function keys, and scientific calculator keys.

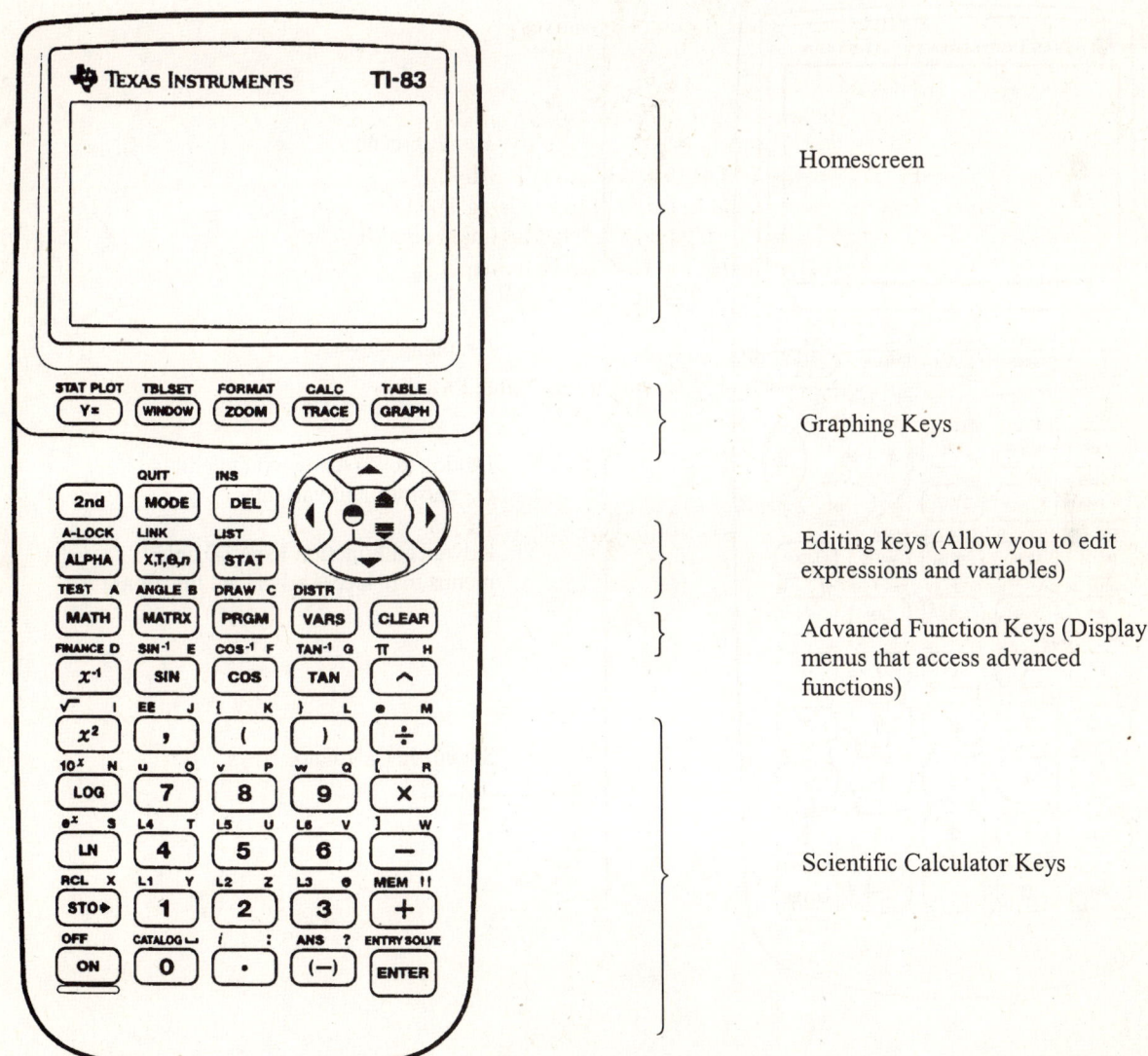

Homescreen

Graphing Keys

Editing keys (Allow you to edit expressions and variables)

Advanced Function Keys (Display menus that access advanced functions)

Scientific Calculator Keys

Operating the TI-83 Plus

TURNING THE CALCULATOR ON AND OFF	[ON]	Turns the calculator on
	[2nd] [ON]	Turns the calculator off
ADJUSTING THE DISPLAY CONTRAST	[2nd] [▲]	Increases the display (darkens the screen)
	[2nd] [▼]	Decreases the contrast (lightens the screen)

Note: If the display begins to dim (especially during calculations), and you must adjust the contrast to 8 or 9 in order to see the screen, batteries are low and you should replace them soon.

The TI-83 Plus keyboard is divided into four zones: graphing keys, editing keys, advanced function keys, and scientific calculator keys.

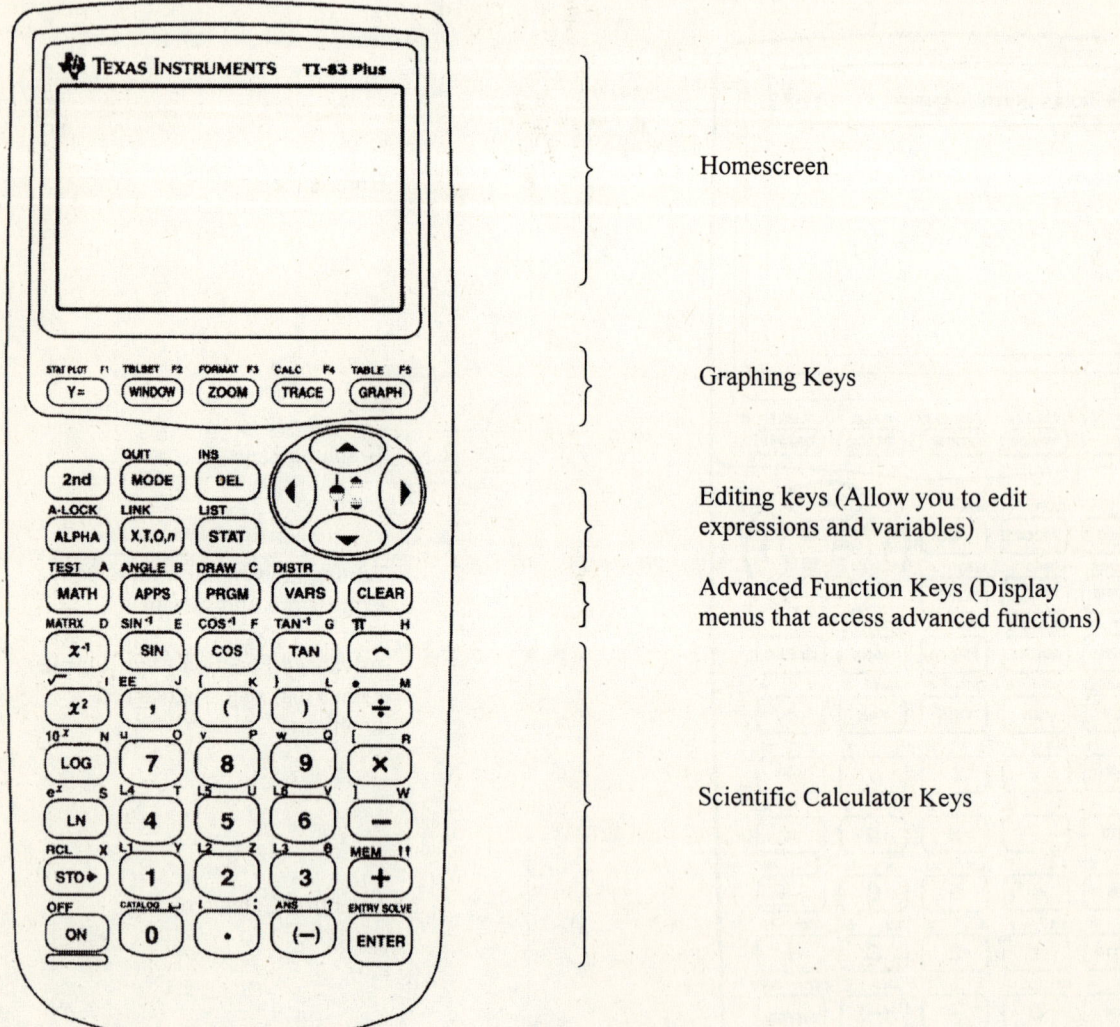

- Homescreen
- Graphing Keys
- Editing keys (Allow you to edit expressions and variables)
- Advanced Function Keys (Display menus that access advanced functions)
- Scientific Calculator Keys

Keystrokes on the TI-83 and TI-83 Plus

[ENTER]	Executes commands or performs a calculation
[2nd]	Pressing the [2nd] key *before* another key accesses the character located above the key and printed in yellow
[ALPHA]	Pressing the [ALPHA] key *before* another key accesses the character located above the key and printed in green
[2nd] [A-LOCK]	Locks in the ALPHA keyboard
[CLEAR]	Pressing [CLEAR] once clears the line
	Pressing [CLEAR] twice clears the screen
[2nd] [QUIT]	Returns to the homescreen
[DEL]	Deletes the character at the cursor
[2nd] [INS]	Inserts characters at the underline cursor
[X,T,Θ,n]	Enters an X in Function Mode, a T in Parametric Mode, a θ in Polar Mode, or an n in Sequence Mode
[STO▶]	Stores a value to a variable
[^]	Raises to an exponent
[2nd] [π]	the number π
[-]	Negative symbol
[MATH] [▶][1]	Computes the absolute value of a number or an expression in parentheses
[2nd] [ENTRY]	Recalls the last entry
[2nd] [:]	Used to enter more than one expression on a line
[2nd] [ANS]	Recalls the most recent answer to a calculation
[x^2]	Squares a number or an expression
[x^{-1}]	Inverse; can be used with a real number or a matrix
[2nd] [$\sqrt{}$]	Computes the square root of number or an expression in parentheses
[2nd] [e^x]	Returns the constant e raised to a power
[ALPHA] [0]	Space
[2nd] [◀]	Moves the cursor to the beginning of an expression
[2nd] [▶]	Moves the cursor to the end of an expression

Special Features of the TI-83 Plus

The TI-83 Plus uses Flash technology, which lets you upgrade to future software versions and to download helpful programs from the TI website, www.ti.com/calc.

Applications can be installed and accessed via the [APPS] key. The Finance application is accessed via this key rather than directly as it is with the TI-83.

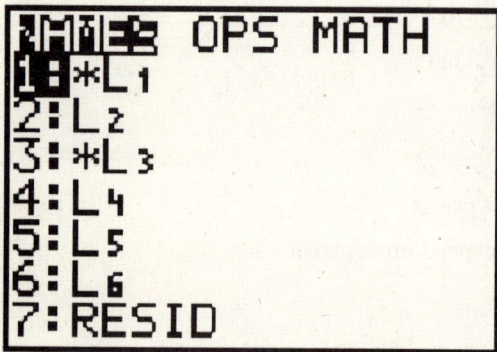

Archiving on the TI-83 Plus allows you to store data, programs, or other variables to user data archive, a protected area of memory separate from RAM, where they cannot be edited or deleted inadvertently.

Archived variables are indicated by asterisks (*) to the left of variable names.

On the TI-83 Plus, MATRX applications are accessed by pressing [2nd] [MATRX] rather than directly as they are with the TI-83.

4

Setting Modes

Mode settings control how the TI-83 and TI-83 Plus displays and interprets numbers and graphs.

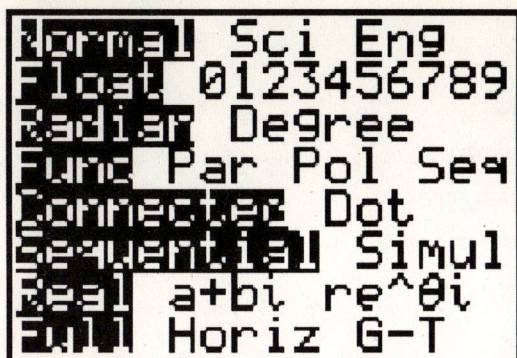

Numeric display format
Number of decimal places
Unit of angle measure
Type of graphing
Whether to connect graph points
Whether to plot simultaneously
Real, rectangular complex, or polar complex
Full, two split-screen modes

MATH Operations

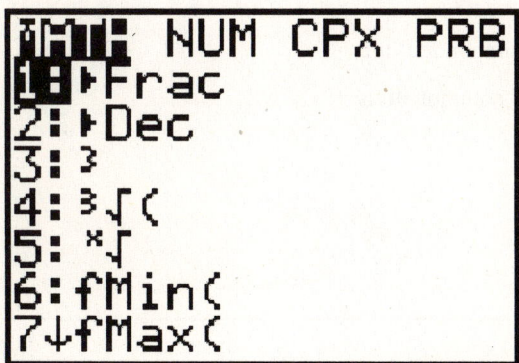

Displays the answer as a fraction
Displays the answer as a decimal
Calculates the cube
Calculates the cube root
Calculates the xth root
Finds the minimum of a function
Finds the maximum of a function

Computes the numerical derivative
Computes the function integral
Displays the equation solver

MATH NUM (Number) Operations

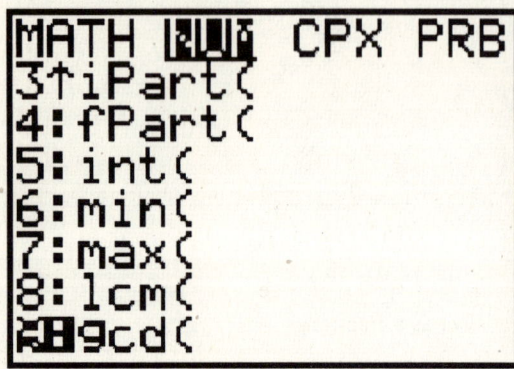

Absolute value
Round
Integer part
Fractional part
Greatest integer
Minimum value
Maximum value

Least common multiple
Greatest common divisor

MATH CPX (Complex) Operations

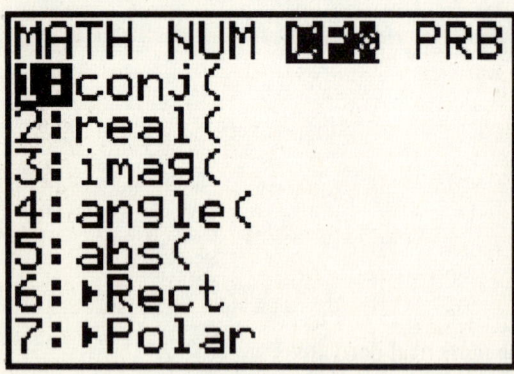

Returns the complex conjugate
Returns the real part
Returns the imaginary part
Returns the polar angle
Returns the magnitude (modulus)
Displays the result in rectangular form
Displays the result in polar form

MATH PRB (Probability) Operations

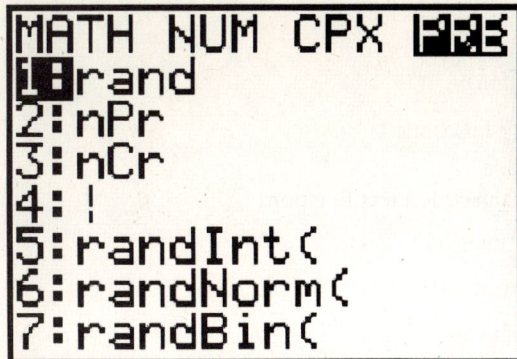

Random-number generator
Number of permutations
Number of combinations
Factorial
Random-integer generator
Random number from Normal distribution
Random number from Binomial distribution

Y= Editor

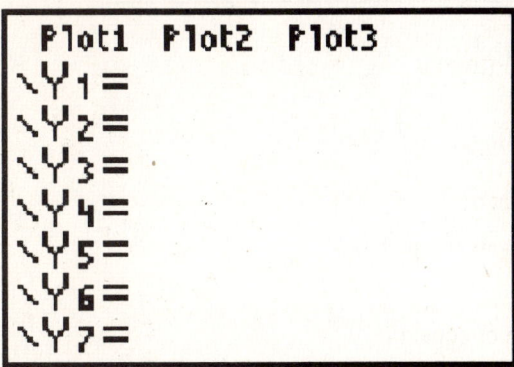

Up to 10 functions can be stored to the function variables Y_1 through Y_9, and Y_0. One or more functions can be graphed at once.

VARS Menu

X/Y, T/θ, and U/V/W variables
ZX/ZY, ZT/Zθ, and ZU variables
GRAPH DATABASE variables
PICTURE variables
XY, Σ, EQ, TEST, and PTS variables
TABLE variables
STRING variables

VARS Y-VARS Menus

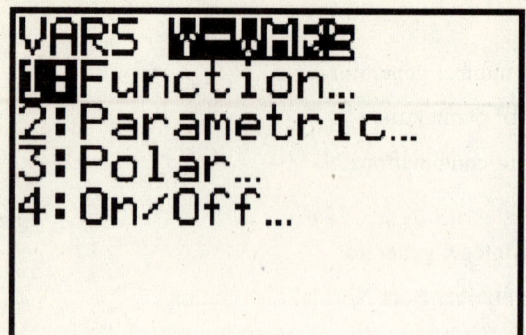

Y_n functions

X_{nT}, Y_{nT} functions

r_n functions

Lets you select/deselect functions

TEST Menu

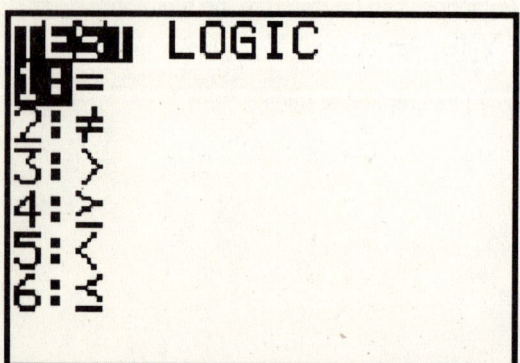

Returns 1 (true) if:

Equal

Not equal

Greater than

Greater than or equal to

Less than

Less than or equal to

TEST LOGIC Menu

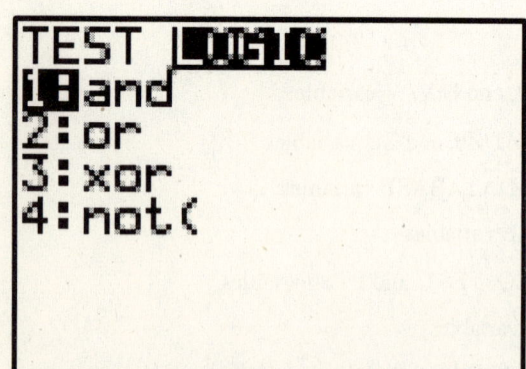

Returns 1 (true) if:

Both values are nonzero (true)

At least one value is nonzero (true)

Only one value is zero (false)

The value is zero (true)

Operating the TI-82

TURNING THE CALCULATOR ON AND OFF
[ON] Turns the calculator on
[2nd] [ON] Turns the calculator off

ADJUSTING THE DISPLAY CONTRAST
[2nd] [▲] Increases the display (darkens the screen)
[2nd] [▼] Decreases the contrast (lightens the screen)

Note: If the display begins to dim (especially during calculations), and you must adjust the contrast to 8 or 9 in order to see the screen, batteries are low and you should replace them soon.

The TI-83 keyboard is divided into four zones: graphing keys, editing keys, advanced function keys, and scientific calculator keys.

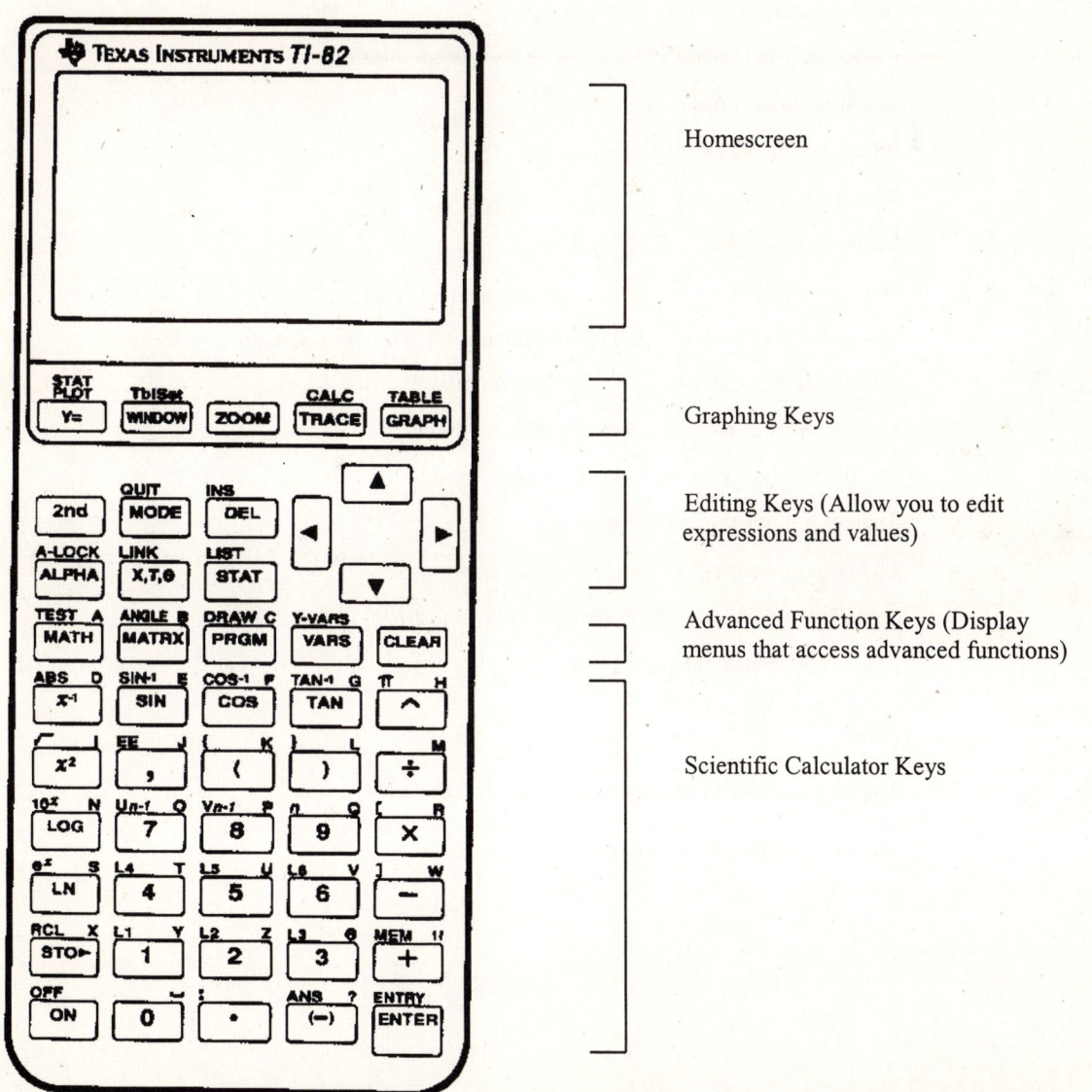

- Homescreen
- Graphing Keys
- Editing Keys (Allow you to edit expressions and values)
- Advanced Function Keys (Display menus that access advanced functions)
- Scientific Calculator Keys

Keystrokes on the TI-82

[ENTER]	Executes commands or performs a calculation
[2nd]	Pressing the [2nd] key *before* another key accesses the character located above the key and printed in yellow
[ALPHA]	Pressing the [ALPHA] key *before* another key accesses the character located above the key and printed in green
[2nd] [A-LOCK]	Locks in the ALPHA keyboard
[CLEAR]	Pressing [CLEAR] once clears the line
	Pressing [CLEAR] twice clears the screen
[2nd] [QUIT]	Returns to the homescreen
[DEL]	Deletes the character at the cursor
[2nd] [INS]	Inserts characters at the underline cursor
[X,T,θ]	Enters an X in Function Mode, a T in Parametric Mode, or a θ in Polar Mode
[STO▶]	Stores a value to a variable
[^]	Raises to an exponent
[2nd] [π]	the number π
[−]	Negative symbol
[2nd] [ABS]	Computes the absolute value of a number or an expression in parentheses
[2nd] [ENTRY]	Recalls the last entry
[2nd] [:]	Used to enter more than one expression on a line
[2nd] [ANS]	Recalls the most recent answer to a calculation
[x^2]	Squares a number or an expression
[x^{-1}]	Inverse; can be used with a real number or a matrix
[2nd] [√]	Computes the square root of number or an expression in parentheses
[2nd] [e^x]	Returns the constant *e* raised to a power
[ALPHA] [0]	Space
[2nd] [◀]	Moves the cursor to the beginning of an expression
[2nd] [▶]	Moves the cursor to the end of an expression

Setting Modes

Mode settings control how the TI-82 displays and interprets numbers and graphs.

Numeric display format
Number of decimal places
Unit of angle measure
Type of graphing
Whether to connect graph points
Whether to plot simultaneously
Full or split-screen mode

MATH Operations

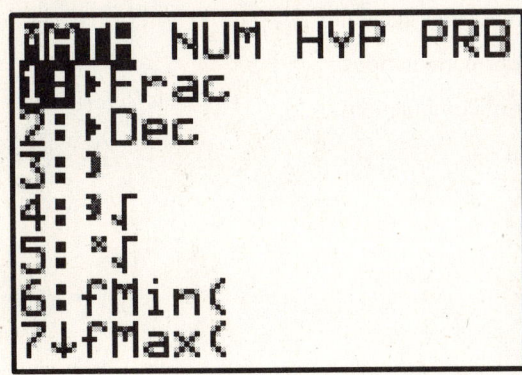

Displays the answer as a fraction
Displays the answer as a decimal
Calculates the cube
Calculates the cube root
Calculates the xth root
Finds the minimum of a function
Finds the maximum of a function

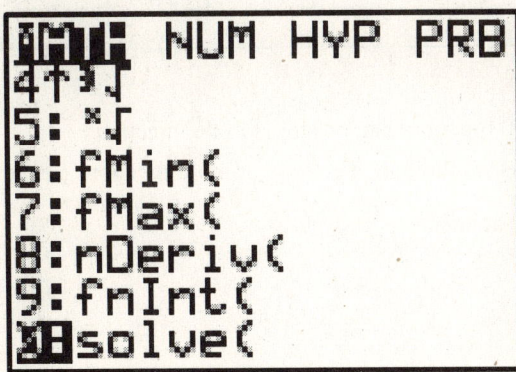

Computes the numerical derivative
Computes the function integral
Computes the solution of a function

MATH NUM (Number) Operations

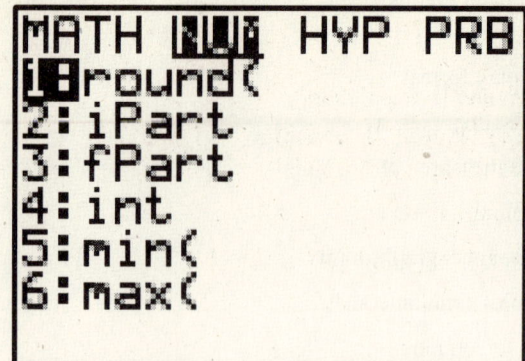

Round
Integer part
Fractional part
Greatest integer
Minimum value
Maximum value

MATH PRB (Probability) Operations

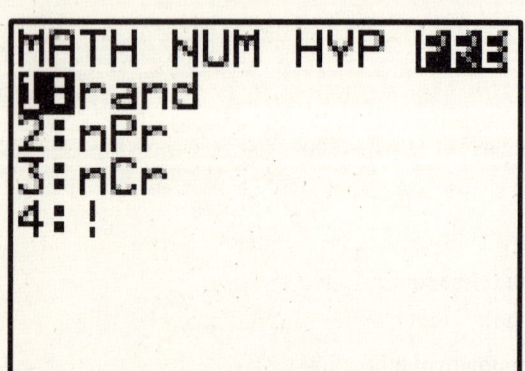

Random-number generator
Number of permutations
Number of combinations
Factorial

Y= Editor

Up to 10 functions can be stored to the function variables Y_1 through Y_8. One or more functions can be graphed at once.

VARS Menu

X/Y, T/θ, and U/V variables
ZX/ZY, ZT/Zθ, and ZU variables
GRAPH DATABASE variables
PICTURE variables
XY, Σ, EQ, BOX, and PTS variables
TABLE variables

2nd [Y-VARS] Menus

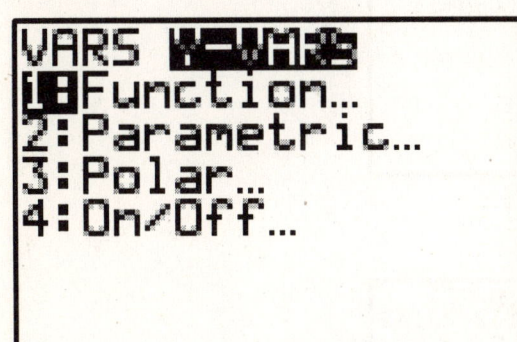

Y_n functions
X_{nT}, Y_{nT} functions
r_n functions
Lets you select/deselect functions

TEST Menu

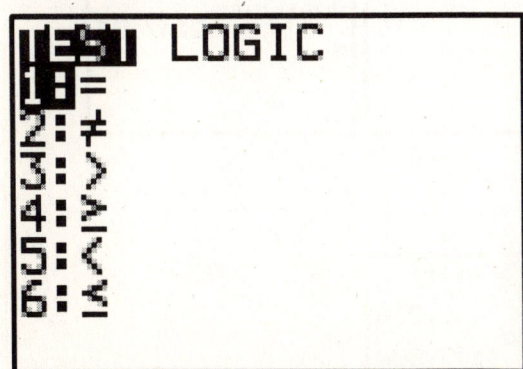

Returns 1 (true) if:
Equal
Not equal
Greater than
Greater than or equal to
Less than
Less than or equal to

I. CALCULATIONS WITH THE TI-83, TI-83 Plus, and TI-82

CAUTION! The negative sign (-) and the subtraction sign − are different. Use the − sign for subtraction and the (-) sign to write negative numbers.

CALCULATIONS	EXAMPLES
Because the TI-83, TI-83 Plus, and the TI-82 use standard algebraic order when evaluating arithmetic expressions, the expression can be entered as it appears. Working outwards from inner parentheses, calculations are performed from left to right. Powers and roots are evaluated first, followed by multiplications and divisions, and then additions and subtractions.	Calculate $-4(9-8)+(-7)(2)^3$ ``` -4(9-8)+(-7)(2)^ 3 -60 ```

If the numerator or denominator of a fraction contains more than one operation, it should be enclosed in parentheses when entering it into the calculator.
 	
Note: To preserve the order of operations when calculations involve fractions, enter the fractions in parentheses.
 	
Decimal answers will normally appear if the answers are not integers. If an answer is a rational number, its fractional form can be found by pressing MATH 1: Frac and pressing ENTER.
	Calculate $\dfrac{10-6}{4-4(3)}$
	
	``
	(10-6)/(4-4*3)
	               -.5
	``
	
	Calculate $\dfrac{3}{7}+\dfrac{19}{27}$
	
	``
	(3/7)+(19/27)
	        1.132275132
	``

When entering an expression to be calculated, be careful to enclose expressions in parentheses as needed.	```
	MATH NUM CPX PRB
	1:▶Frac
	2:▶Dec
	3:³
	4:³√(
	5:ˣ√
	6:fMin(
	7↓fMax(
	``` ```
	(3/7)+(19/27)
	        1.132275132
	Ans▶Frac
	           214/189
	``
	
	Calculate $3 \div 4 + \dfrac{2^5}{4^2} + \left(-\dfrac{5}{4}\right)^{-3}$
	
	``
	3/4+(2^5/4²)+(-5
	/4)^(-3)
	             2.238
	Ans▶Frac
	          1119/500
	``

CALCULATIONS WITH RADICALS AND RATIONAL EXPONENTS	EXAMPLES
Square roots can be evaluated by using the $\sqrt{}$ key, when the expression is defined in the set of real numbers. If the expression is undefined, an error message appears.	Calculate $\sqrt{289}$ $\sqrt{-289}$ 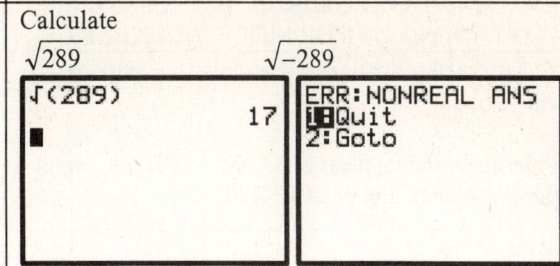
Cube roots can be evaluated by pressing MATH 4: $\sqrt[3]{}$.	Calculate $\sqrt[3]{-3375}$ 
Roots of other orders can be evaluated by entering the index and then pressing MATH 5: $\sqrt[x]{}$	Calculate $\sqrt[4]{4096}$ 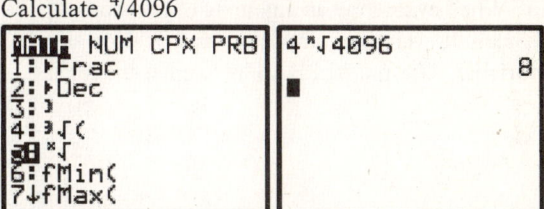
Recall that roots can be converted to fractional exponents using $\sqrt[n]{a} = a^{(1/n)}$ and $\sqrt[n]{a^m} = a^{(m/n)}$.	Another way to calculate $\sqrt[4]{4096}$ 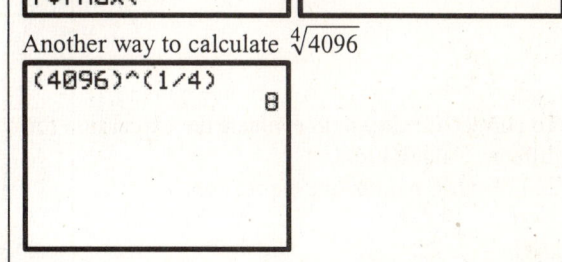
On the TI-83 and TI-83 Plus, expressions containing rational exponents can be evaluated. On the TI-82, some expressions may have to be rewritten using the property $a^{m/n} = \left(a^{1/n}\right)^m$.	Calculate $(-64)^{2/3}$ (On the TI-82): 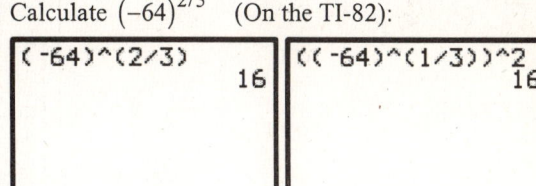
If the result of a computation is an irrational number, only the decimal approximation of this irrational number will be shown. Pressing MATH, 1: Frac will not give a fraction; it will give the same decimal.	Calculate $\dfrac{\sqrt{18}}{\sqrt{3}}$ 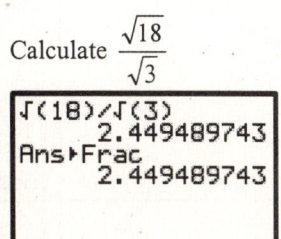

II. EVALUATING ALGEBRAIC EXPRESSIONS

EVALUATING ALGEBRAIC EXPRESSIONS CONTAINING ONE OR MORE VARIABLES	EXAMPLE
| 1. To evaluate an algebraic expression containing one variable:<br><br>Enter the x-value, press STO, X, ALPHA, : , enter the expression, and press ENTER.<br><br><br>2. To evaluate an algebraic expression containing two variables, *x* and *y*:<br><br>Enter the x-value, press STO, X, ALPHA, : , enter the y-value, STO, ALPHA, Y, ALPHA, : , enter the expression, and press ENTER.<br><br>3. When evaluating an algebraic expression containing variables, any letter may be used as a variable. Use the ALPHA key to enter the letters.<br><br><br><br><br><br>To correct an entry or to evaluate the expression for different values, press<br>2nd ENTER and edit the expression. | 1. Evaluate $\dfrac{x+4}{5-x}$ for x = –6.<br>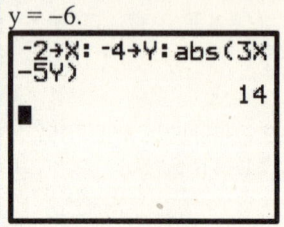<br><br>2. Evaluate $\lvert 3x - 5y \rvert$ for x = –2 and y = –6.<br><br>```
-2→X: -4→Y:abs(3X
-5Y)
                14
```<br><br>3. Find the surface area of a right circular cylinder with r = 5.2 and h = 6.4.<br><br>The formula for the surface area is $S = 2\pi R^2 + 2\pi RH$, so enter<br>$5.2 \to R:6.4 \to H:2\pi R^2 + 2\pi RH$ and press ENTER.<br><br>```
5.2→R:6.4→H:2πR²
+2πRH
         379.0017377
```<br><br>The output (surface area) is 379.002, to three decimal places.<br><br>The surface area of a right circular cylinder for different values of r and h can be found by pressing 2nd ENTER and entering the new values. For r = 1.3 and h = 2.7:<br><br>```
5.2→R:6.4→H:2πR²
+2πRH
         379.0017377
1.3→R:2.7→H:2πR²
+2πRH
          32.6725636
``` |
```

III. GRAPHING EQUATIONS

USING A GRAPHING CALCULATOR TO GRAPH AN EQUATION	EXAMPLES
To graph an equation in the variables x and y:	Graph $2x^2 - 2y = 3$ with a graphing calculator.
1. Solve the equation for y in terms of x.	1. Solving for y gives $-2y = -2x^2 + 3$, so $y = \dfrac{-2x^2}{-2} + \dfrac{3}{-2}$ or $y = x^2 - \dfrac{3}{2}$.
2. Press the Y= key to access the function entry screen and enter the equation into y_1. Use parentheses as needed so that what you have entered agrees with the order of operations. To erase equations from the equation editor, press CLEAR. To leave the equation editor and return to the homescreen, press 2nd QUIT.	2. Press Y= key and enter $y_1 = x^2 - 3/2$. Both screens below will give the graph. 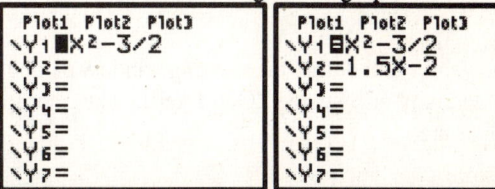
3. Determine an appropriate viewing window. Frequently the standard window (ZOOM 6) is appropriate, but often a decimal or integer viewing window (ZOOM 4 or ZOOM 8) gives a better representation of the graph. Pressing GRAPH or a ZOOM key will activate the graph. ZOOM 8 must be followed by ENTER, but ZOOM 4 and ZOOM 6 do not.	3. The graphs of y = $x^2 - 3/2$ using possible windows are below. ZOOM 6 (standard) ZOOM 4 (decimal) 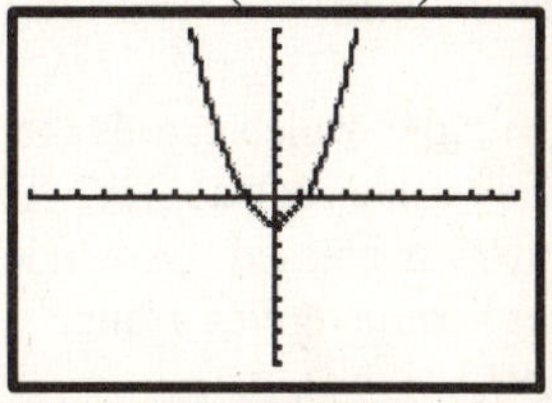 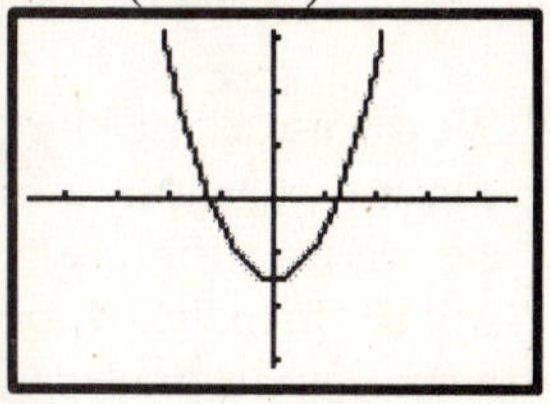 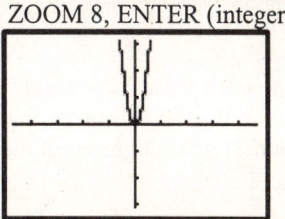 ZOOM 8, ENTER (integer) 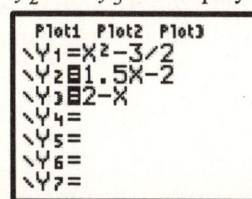
4. All equations in the equation editor that have their "=" signs highlighted (dark) will have their graphs shown when GRAPH is pressed. If the "=" sign of an equation in the equation editor is not highlighted, the equation will remain, but its graph is "turned off" and will not appear when GRAPH is pressed. The graph is "turned on" by repeating the process.)	4. Equation y_1 is turned off. Equation y_2 and y_3 are turned on. Only the graphs of y_2 and y_3 are displayed. 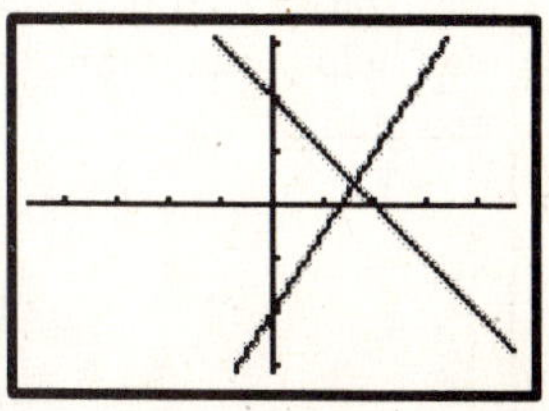

VIEWING WINDOWS	EXAMPLE
With a TI-83, TI-83 Plus, and TI-82, as with a graph plotted by hand, the appearance of the graph is determined by the part of the graph we are viewing. The viewing window determines how a given graph appears in the same way that different camera lenses show different views of an event. The values that define the viewing window can be set individually or by using ZOOM keys. The important values are: x-min: the smallest value on the x-axis x-max: the largest value on the x-axis x-scale: spacing for tics on the x-axis y-min: the smallest value on the y-axis y-max: the largest value on the y-axis y-scale: spacing for tics on the y-axis We can use a "friendly" window, which causes the cursor to change by a "nice" value such as .1, .2, 1, etc. when a right or left arrow is pressed. A window will be "friendly" if *xmax – xmin* gives a "nice" number when divided by 94. ZOOM, 4 automatically gives *xmin* = -4.7 and *xmax* = 4.7, so *xmax – xmin* = 9.4. Thus each press of the right or left arrow moves the cursor 9.4/94 = .1 units. ZOOM, 8, ENTER gives a movement of 1 unit for each press of an arrow. The window should be set so that the important parts of the graph are shown and the unseen parts are suggested. Such a graph is called **complete**. The values that define the viewing window can be set individually. If necessary, using ZOOM, 3:Zoom Out can help to determine the shape and important parts of the graph.	The graph of $y = x^3 - x$ looks somewhat like a line in the region resulting from ZOOM 8. 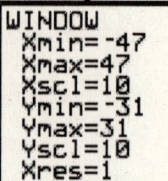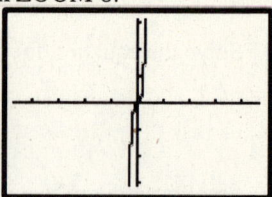 But its shape is defined better in the standard viewing window, accessed by pressing ZOOM 6, giving a window with x-values and y-values between -10 and 10. The graph of $y = x^3 - x$ with ZOOM 6: 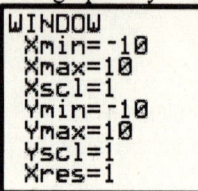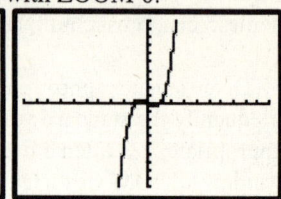 The graph of $y = x^3 - x$ with ZOOM 4: 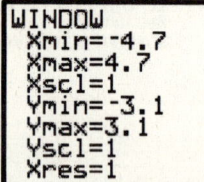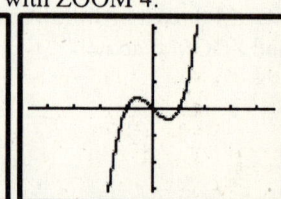 The following window shows the complete graph clearly. 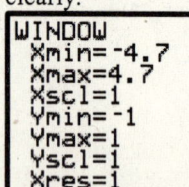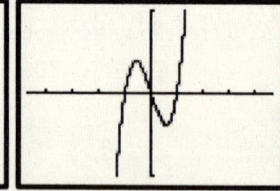 Note: Using *xmin* = -9.4 and *xmax* = 9.4 with *ymin* = -10 and *ymax* = 10 gives a window that is "friendly" and close to the standard window.

FINDING y-VALUES FOR SPECIFIC VALUES OF x	EXAMPLE
To find y-values at selected values of x by using TRACE, VALUE: 1. Press the Y= key to access the function entry screen and enter the right side of the equation. Use parentheses as needed so that what you have entered agrees with the order of operations. 2. Set the window so that it contains the x-value whose y-value we seek. 3. Press GRAPH. OR Do the following, which we will call TRACE, VALUE: 4. ON THE TI-83 and TI-83 PLUS: Press TRACE and then enter the selected x-value followed by ENTER. The cursor will move to the selected value and give the resulting y-value if the selected x-value is in the window. If the selected x-value is not in the window, Err: INVALID occurs. If the x-value is in the window, the y-value will occur even if it is not in the window. ON THE TI-82 AND ON THE TI-83 AND TI-83 PLUS: 5. Press 2nd calculate, 1:(value), ENTER, enter the x-value, and press ENTER. The corresponding y-value will be displayed if the selected x-value is in the window.	Find y when x is 2, 6, and -1 for the equation y = 5x − 1. 1. Enter $y_1 = 5x - 1$ 2. Set the window with *xmin* = −10 and *xmax* = 10, *ymin* = −10 and *ymax* = 10. 3. 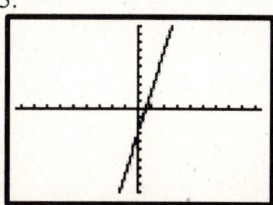 4. Press TRACE, enter the value 2, getting y = 9. 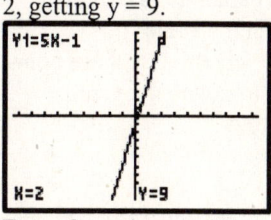 Enter 6, getting y = 29 Enter -1, getting y = -6 5. Press TRACE, enter 6, and press ENTER, getting y = 29. 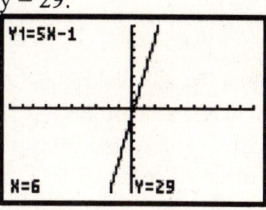 Enter 2, get y = 9. Enter −1, get y = −6. 6. Press 2nd calculate, 1:(value), ENTER, enter −1, and press ENTER, getting y = −6. 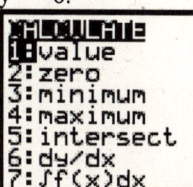

GRAPHING EQUATIONS ON PAPER	EXAMPLE
To graph an equation in the variables x and y: 1. Solve the equation for y in terms of x. 2. Press the Y= key to access the function entry screen and enter the equation. Use parentheses as needed so that what is entered agrees with the order of operations. 3. Determine an appropriate viewing that gives a complete graph. Press GRAPH. 4. To sketch the graph on paper, use TABLE or TRACE, VALUE to get x and y values of representative points on the graph. Use these points and the shape of the graph to sketch the graph. USING THE TABLE 1. a. To prepare the table, press 2nd Table Set, enter an initial x-value in a table (Tblmin), and enter the change (Δ Tbl) in the x-value we want in the table. b. If we change Independent variable to Ask, we may enter any value of x we choose and get the corresponding y-value. 2. Enter 2nd TABLE to get a list of x-values and the corresponding y-values. The value of the function at the given value of x can be read from the table. OR simply enter the x-value if Independent variable is set to Ask. 3. If Independent variable is at Auto, use the up or down arrow to find the x-values where the function is to be evaluated.	Graph $2x - 3y = 12$. 1. Solving for y gives $-3y = -2x + 12$ $$y = \frac{-2x}{-3} + \frac{12}{-3} \text{ or } y = \frac{2x}{3} - 4$$ 2. Enter $y_1 = 2x/3 - 4$ NOTE: Division is preserved without using parentheses in this case. 3. Graphing the equation with *xmin* = -9.4 and *xmax* = 9.4, and with *ymin* = -10 and 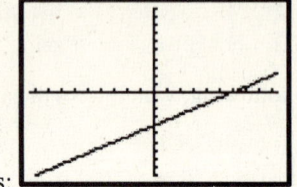 *ymax* = 10 gives: 4. TRACE, VALUE gives coordinates of points. 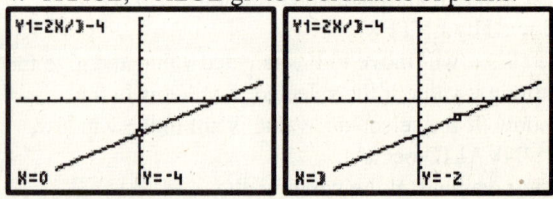 1. Setting Tblmin = 0 and Δ Tbl = 1 gives the table below. 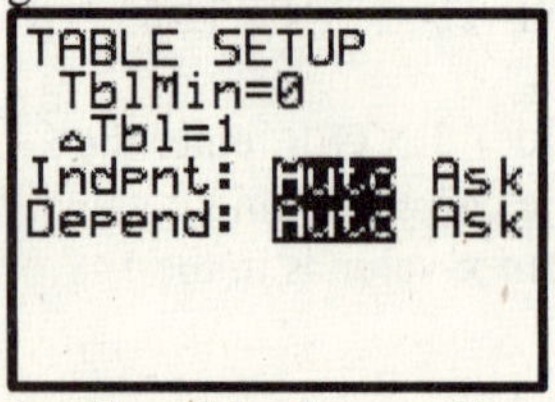  2. The table shows a list of values of x and the corresponding values of y. The value of y when x is 3 is -2. 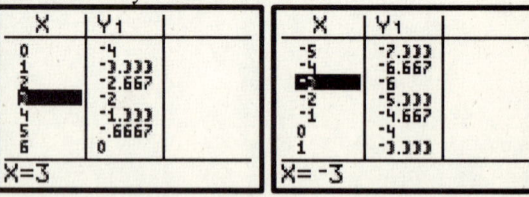 3. Pressing the up arrow gives values less than 0 and the corresponding y-values. The value corresponding to -3 is -6.

20

USING A GRAPHING CALCULATOR TO GRAPH EQUATIONS CONTAINING y^2	EXAMPLE
To use a graphing calculator to graph an equation containing y^2: 1. Solve the equation for y using the property that $y^2 = a$ if and only if $y = \pm\sqrt{a}$. 2. Use the Y= key to enter each of the solutions for y. 3. Determine an appropriate viewing window and graph the equation. 4. The graph will be in correct proportion (square) if xmax − xmin is some multiple of 94 while ymax − ymin is that same multiple of 62. ZOOM 4 and ZOOM 8 are square windows, and ZOOM 5 changes the y-values of the window so that the window is square.	Graph the circle with equation $x^2 + y^2 = 49$ 1. Solve for y. $y^2 = 49 - x^2$ $y = \pm\sqrt{49 - x^2}$ 2. Enter $y_1 = \sqrt{(49 - x^2)}$ and $y_2 = -\sqrt{(49 - x^2)}$. 3. The standard window is appropriate but does not produce a graph which appears to be a circle. 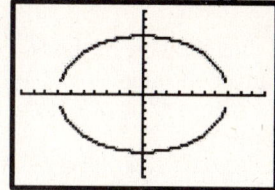 4. Using a SQUARE window (ZOOM 5) will correct this dilemma. 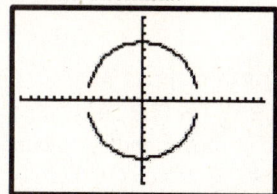
ADDITIONAL EXAMPLE	Graph $x^2 - 4y^2 = 16$ on a square window. Solving for y gives two equations $y^2 = \dfrac{16 - x^2}{-4} = \dfrac{x^2 - 16}{4}$ $y = \dfrac{\sqrt{x^2 - 16}}{2}$, $y = -\dfrac{\sqrt{x^2 - 16}}{2}$ Entering the equations as y_1 and y_2 gives the graph of the relation. 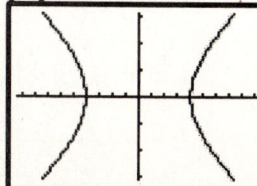

IV. EVALUATING FUNCTIONS

If y is a function of x, then the y-coordinate of the graph at a given value of x is the functional value.

EVALUATING FUNCTIONS WITH TRACE, VALUE	EXAMPLE
To evaluate functions at selected values of x by using TRACE:	Evaluate f(2), f(6), and f(−1) if f(x) = 5x − 1.
1. Use the Y= key to store y_1 = f(x).	1. Enter y_1 = 5x − 1
2. Graph using an appropriate viewing window that gives a complete graph.	2.
3. Use TRACE with one of the following methods. The resulting y-value is the function value. a. ON THE TI-83 and TI-83 PLUS: Press TRACE and then enter the selected x-value followed by ENTER. The corresponding y-value will be displayed if the selected x-value is in the window.	3. a. Press TRACE, enter 2, getting y = 9. Thus f(2) = 9.
ON THE TI-82 OR THE TI-83 OR TI-83 PLUS: b. Press 2nd calculate, 1:(value), ENTER, enter the x-value, and press ENTER. The corresponding y-value will be displayed if the selected x-value is in the window.	b. Press 2nd calculate, 1:(value), ENTER, enter −1, and press ENTER, getting y = −6. Thus f(−1) = −6.

22

We can also evaluate functions by means other than TRACE. Some alternate ways follow.

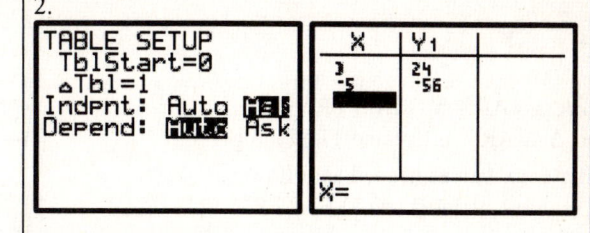

EVALUATING A FUNCTION WITH TABLE	EXAMPLE
To evaluate a function with a table:	Evaluate $y = -x^2 + 8x + 9$ when $x = 3$ and when $x = -5$.
1. Enter the function with the Y= key.	1. Enter $y_1 = -x^2 + 8x + 9$.
2. To find f(x) for specific values of x in the table, press 2nd Table Set, move the cursor to Ask opposite Indpnt:, and press ENTER. Then press 2nd TABLE and enter the specific values.	2.
ALTERNATE METHOD: 3. Press 2nd Table Set, enter an initial x-value in a table (Tblmin), and enter the desired change (Δ Tbl) in the x-value in the table.	3. Setting Tblmin = 0 and Δ Tbl = 1 gives the table below. 
4. Enter 2nd TABLE to get a list of x-values and the corresponding y-values. The value of the function at the given value of x can be read from the table.	4. The table shows a list of values of x and the corresponding values of y. The value of y when x is 3 is 24. 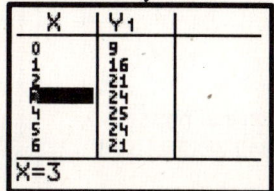 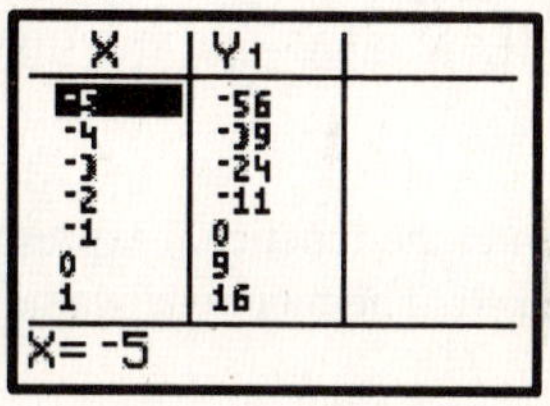
5. Use the up or down arrow to find the x-values where the function is to be evaluated.	5. Pressing the up arrow gives values less than 0 and the corresponding y-values. The value corresponding to –5 is –56.

EVALUATING FUNCTIONS WITH y-VARS	EXAMPLE
To evaluate the function f at one or more values of x: 1. Use the Y= key to store $y_1 = f(x)$. Press 2nd QUIT. 2. Press VARS, Y-VARS 1,1 (2nd, Y-VARS, 1,1 on the TI-82) to get y_1. Enter the x-values needed as follows: y_1 ({value 1, value 2, etc.}) ENTER. Values of the function will be displayed.	Given the function $f(x) = -16x^2 + 20x - 2$, find $f(4), f(1.45), f(-2), f(-8.4)$. 1. Enter $y_1 = -16x^2 + 20x - 2$ in y_1: 2. Enter y_1 ({4,1.45,-2,84}) The display gives the values {-178, -6.64, -106, -111,218}.

EVALUATING FUNCTIONS OF SEVERAL VARIABLES	EXAMPLES
To evaluate a function of two variables: 1. Enter the x-value, press STO, X, ENTER. Enter the y-value, press STO, ALPHA Y (above 1), ENTER. 2. Enter the functional expression and press ENTER. The functional value will be displayed. 3. To evaluate the function for different values of the variables, enter new values of the variables and press 2nd ENTER to repeat the functional expression.	If $z = f(x,y) = 2x + 5y$, find $f(3, 5)$. 1. Enter 3 for x and 5 for y, using STO: 2. Entering 2x + 5y and pressing ENTER gives $f(3, 5)$ 3. To evaluate the same function at $x = -3, y = 4$, enter the new values for x and y, and use 2nd ENTER to find the functional value.

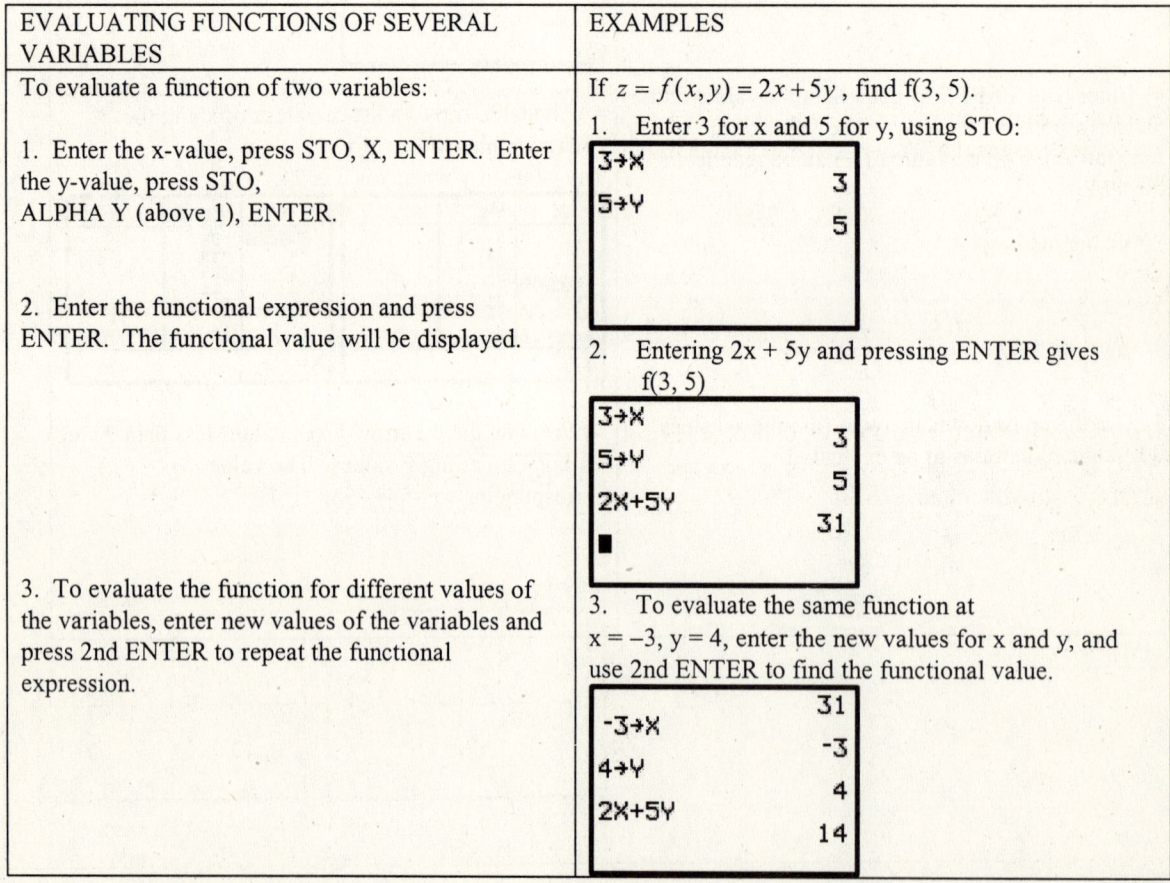

V. DOMAINS AND RANGES OF FUNCTIONS; COMBINATIONS OF FUNCTIONS

The graphs of functions can be used to determine or to verify their domains and ranges.
  Domain: set of all x-values for which a function is defined
  Range: set of all y-values resulting from these x-values

FINDING OR VERIFYING DOMAINS AND RANGES OF FUNCTIONS	EXAMPLES
To visually find the domain and range of a function: 1. Graph the function with a window that shows all the important parts and suggests where unseen parts are located. Such a graph is called complete. 2. Visually determine if the graph is defined for the set of all x-values (all real numbers) or for some subset of the real numbers. This set is the domain of the function. 3. Visually determine if the y-values on the graph form the set of all real numbers or some subset of the real numbers. This set is the range of the function. If a function contains a denominator, values of x that make the denominator 0 are not in the domain of the fraction. Graphing the function with a window that contains these values of x shows that the graph is not defined for them. If a function contains a square root radical, values of x that give negative values inside the radical are not in the domain of the function.	Find the domains and ranges of the following functions. $f(x) = 3x-8$ $f(x) = x^2+2$ 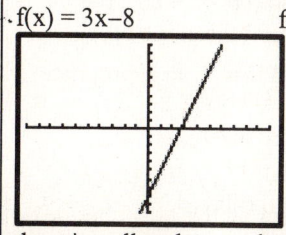 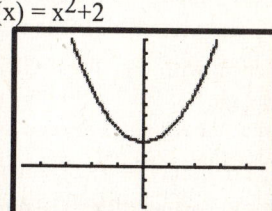 domain: all reals domain: all reals range: all reals range: $\{y \mid y \geq 2\}$ $f(x) = -x^3+5$ $f(x) = -0.4x^4 + 8$ 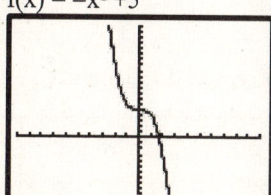 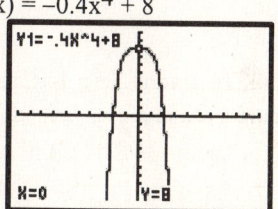 domain: all reals domain: all reals range: all reals range: $\{y \mid y \leq 8\}$ $f(x) = \dfrac{4}{x+2}$ 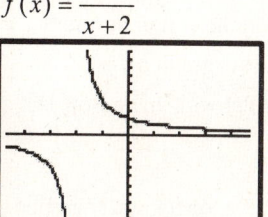 domain: reals except –2 range: reals except 0 $y = \sqrt{2+x}$ $g(x) = \sqrt{4-x^2}$ 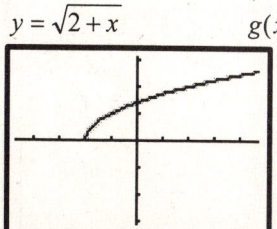 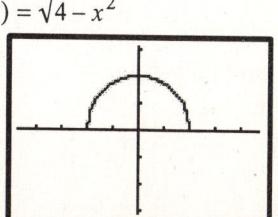 domain: $\{x \mid x \geq -2\}$ domain: $\{x \mid -2 \leq x \leq 2\}$ range: $\{y \mid y \geq 0\}$ range: $\{y \mid 0 \leq y \leq 2\}$

COMBINATIONS OF FUNCTIONS	EXAMPLES
To find the graphs of combinations of two functions f(x) and g(x):	If $f(x) = 4x - 8$ and $g(x) = x^2$, find the following:

To find the graphs of combinations of two functions f(x) and g(x):

1. Enter f(x) as y_1 and g(x) as y_2 under the Y= menu.

2. To graph (f + g)(x), enter $y_1 + y_2$ as y_3 under the Y= menu. Place the cursor on the = sign beside y_1 and press ENTER to turn off the graph of y_1. Repeat with y_2. Press GRAPH with an appropriate window.

3. To graph (f − g)(x), enter $y_1 - y_2$ as y_3 under the Y= menu. Place the cursor on the = sign beside y_1 and press ENTER to turn off the graph of y_1. Repeat with y_2. Press GRAPH.

4. To graph ($f*g$)(x), enter y_1*y_2 as y_3 under the Y= menu. Place the cursor on the = sign beside y_1 and press ENTER to turn off the graph of y_1. Repeat with y_2. Press GRAPH.

5. To graph (f/g)(x), enter y_1/y_2 as y_3 under the Y= menu. Place the cursor on the = sign beside y_1 and press ENTER to turn off the graph of y_1. Repeat with y_2. Press GRAPH with an appropriate window.

6. To evaluate f + g, f − g, f * g, or f/g at a specified value of x, enter y_1, the value of x enclosed in parentheses, the operation to be performed, y_2, the value of x enclosed in parentheses, and press ENTER. Or, if the combination of functions is entered as y_3, enter y_3 and the value of x enclosed in parentheses.

Examples:

1. 2. Graph (f + g)(x)
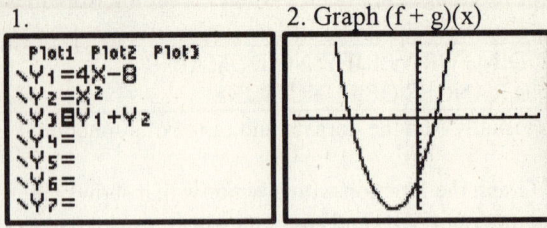

3. Graph (f − g)(x)
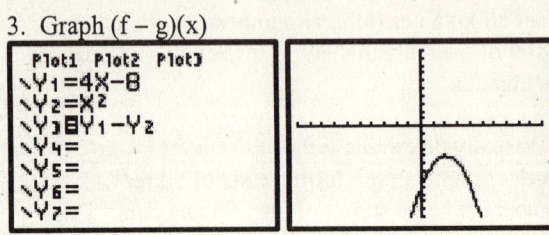

4. Graph (f*g)(x)
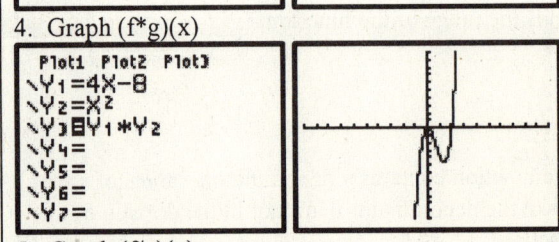

5. Graph (f/g)(x)
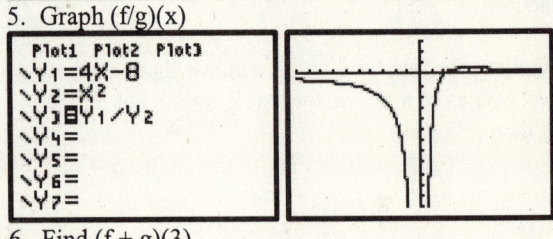

6. Find (f + g)(3).
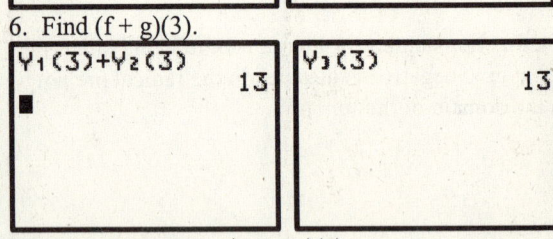

NOTE that entering $(Y_1 + Y_2)(3)$ does not produce the correct result.

COMPOSITION OF FUNCTIONS	EXAMPLES
To graph the composition of two functions f(x) and g(x):	If $f(x) = 4x - 8$ and $g(x) = x^2$, find the following:
1. Enter f(x) as y_1 and g(x) as y_2 under the Y= menu.	1. Enter the two functions. 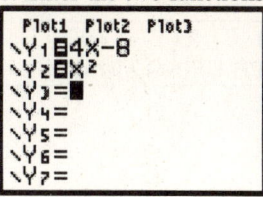
2. To graph $(f \circ g)(x)$, enter $y_1(y_2)$ as y_3 under the Y= menu. Place the cursor on the = sign beside y_1 and press ENTER to turn off the graph of y_1. Repeat with y_2. Press GRAPH with an appropriate window.	2. Graph $(f \circ g)(x) = f(g(x))$ 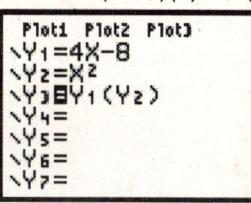 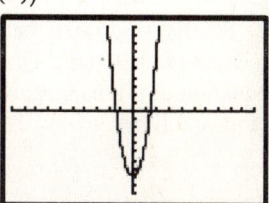
3. To graph $(g \circ f)(x)$, enter $y_2(y_1)$ as y_4 under the Y= menu. Turn off the graphs of y_1, y_2, and y_3. Press GRAPH with an appropriate window.	3. Graph $(g \circ f)(x) = g(f(x))$ 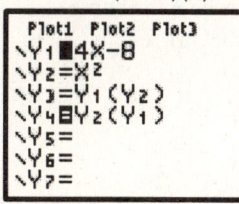 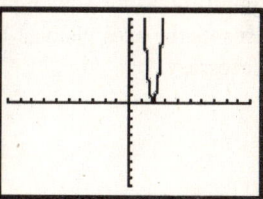
4. To evaluate $(f \circ g)(x)$ at a specified value of x, enter $y_1(y_2)$ and the value of x enclosed in parentheses, and press ENTER. Or, if the combination of functions is entered as y_3, enter y_3 and the value of x enclosed in parentheses.	4. Evaluate $(f \circ g)(-5) = f(g(-5))$ 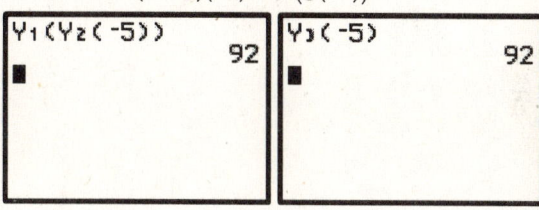
5. To evaluate $(g \circ f)(x)$ at a specified value of x, enter $y_2(y_1)$ and the value of x enclosed in parentheses, and press ENTER. Or, if the combination of functions is entered as y_4, enter y_4 and the value of x enclosed in parentheses.	5. Evaluate $(g \circ f)(-5) = g(f(-5))$ 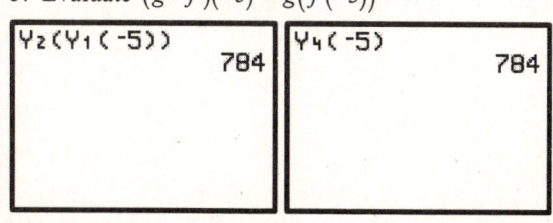

VI. FINDING INTERCEPTS OF GRAPHS

As we have seen, TRACE allows us to find a specific point on the graph. Thus TRACE can be used to solve a number of important problems in algebra. For example, it can be used to find the x- and y-intercepts of a graph.

FINDING OR APPROXIMATING Y- AND X-INTERCEPTS OF A GRAPH USING TRACE	EXAMPLE
To find the y-intercepts and x-intercepts of a graph:	Find the x- and y- intercepts of the graph of $y = -x^2 + 8x + 9$.
1. Solve the equation for y. Enter the equation with the Y= key.	1. Enter $y = -x^2 + 8x + 9$.
2. Set the window so that the intercepts to be located can be seen. The graph of an *n*th degree equation (n is a positive integer) crosses the x-axis at most *n* times.	2. Set the window with *xmin* = −9.4 and *xmax* = 9.4, *ymin* = −10 and *ymax* = 10.
3. Graph the equation. Change the y-values in the window so that the graph is one curve with points where the curve crosses the axes visible. Use ZOOM, Zoom Out if necessary.	3. Graphs (before and after change).
4. Press TRACE and enter the value 0. The resulting y value is the y-intercept of the graph.	4. Press TRACE, enter 0, giving the y-intercept $y = 9$.
5. Press TRACE and use the right and left arrows to move the cursor to values of x that give y = 0. These are the x-intercepts of the graph.	5. Press TRACE, move with arrow to $x = -1$, which gives $y = 0$ (and so $x = -1$ is an x-intercept). Tracing to $x = 9$ gives $y = 0$ (and so $x = 9$ is an x-intercept).

Methods other than TRACE can be used to find the x-intercepts of a graph.

USING CALC, ZERO TO FIND THE X-INTERCEPTS OF A GRAPH	EXAMPLE
To find the x-intercepts of the graph of an equation by using CALC, ZERO: 1. Solve the equation for y. Using the Y= key, enter the equation. 2. Graph the equation with an appropriate window, and note that the equation will intersect the x-axis where y = 0; that is, when x is a solution to the original equation. Set the window so that all points where the graph crosses the x-axis are visible. (Using ZOOM OUT can help check that all such points on the graph are present.) 3. To find the point(s) where the graph crosses the x-axis and the function has zeros, use 2nd CALC, 2 (ZERO). Answer the question "*left bound?*" with ENTER after moving the cursor close to and to the left of an x-intercept. Answer the question "*right bound?*" with ENTER after moving the cursor close to and to the right of this x-intercept. To the question "*guess?*" press ENTER. The coordinates of the x-intercept will be displayed. Repeat to get all x-intercepts. 4. The steps on the TI-82 are identical except 2nd CALC 2 shows the word ROOT, "left bound" is written as "lower bound," and "right bound" is written as "upper bound."	Find the x-intercepts of the graph of $2x^2 - 9x - y = 11$. 1. $y = 2x^2 - 9x - 11$ 2. Because this is a quadratic function, it could have two x-intercepts. Pick x-values that give a friendly window including the x-intercepts. In this case, using -9.4 to 9.4 shows the x-intercepts: 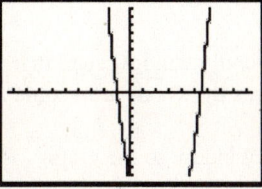 3. Using TRACE shows a solution near x = 5.4. Use 2nd CALC, 2 (ZERO). 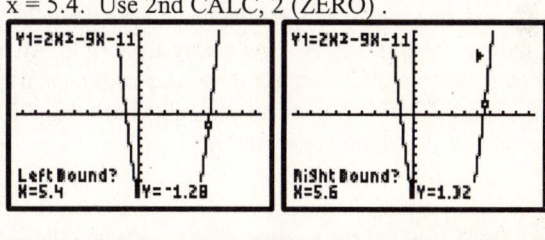 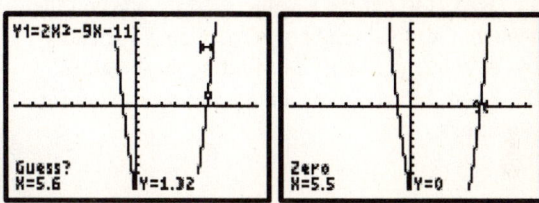 The value of the x-intercept is 5.5. Using TRACE or the steps above gives the second x-intercept as $x = -1$. 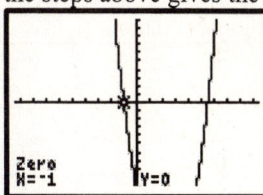

29

VII. SOLVING EQUATIONS

TRACE can be used to find solutions of equations. The equation can be solved by setting one side of the equation equal to zero, graphing y = the nonzero side, and finding the x-intercepts of this graph.

USING TRACE TO FIND OR CHECK SOLUTIONS OF EQUATIONS	EXAMPLE
To solve an equation (approximately) with a graphing calculator:	Solve $x^2 + 11x = -10$.
1. Rewrite the equation with 0 on one side of the equation.	1. $x^2 + 11x + 10 = 0$
2. Using the Y= key, enter the non-zero side of the equation in step 1.	2. $y_1 = x^2 + 11x + 10$
3. Graph the equation with a friendly window such that the points where the graph crosses the x-axis are visible. a. For a linear equation, set the window so the graph crosses the x-axis in one point. b. For a non-linear equation, set the window so that all points where the graph crosses the x-axis are visible. An nth degree equation will intersect the x-axis in at most n points. (Using ZOOM, Zoom Out can help check that all such points on the graph are present.)	3. Because this is a quadratic function, it could have two x-intercepts. Pick x-values that give a friendly window including the x-intercepts. In this case, using x-values from -18.8 to 0 shows the two points where the graph crosses the x-axis.
4. Use TRACE to get the x-value(s) of point(s) on the graph where the y-value(s) are zero. Using a friendly window is frequently helpful in finding these values. ZOOM, Zoom In may give a better approximation for some solutions.	4. Using TRACE shows that y = 0 at x = -10 and at x = -1.
5. The graph will intersect the x-axis where y = 0; that is, where x is a solution to the original equation	5. Thus the solutions to $x^2 + 11x = -10$ are x = -10 and at x = -1.

Methods other than TRACE can be used to solve an equation. The word "zero" means a value of x that makes an expression zero, so an equation can be solved for x by setting one side of the equation equal to zero and using the "zero" method on the TI-83 or TI-83 Plus. The same commands give "roots" (solutions) on the TI-82.

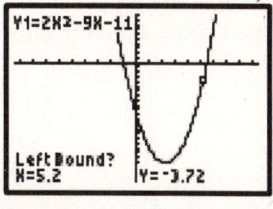

SOLVING EQUATIONS WITH THE ZERO (OR ROOT) METHOD	EXAMPLE
To solve an equation (approximately) with the TI-83 or TI-83 Plus:	Solve $2x^2 - 9x = 11$.
1. Rewrite the equation with 0 on one side of the equation.	1. $2x^2 - 9x - 11 = 0$
2. Using the Y= key, enter the non-zero side of the equation.	2. $y = 2x^2 - 9x - 11$
3. Graph the equation with an appropriate window, and note that the equation will intersect the x-axis where y = 0; that is, when x is a solution to the original equation. Set the window so that all points where the graph crosses the x-axis are visible. (Using ZOOM OUT can help check that all such points on the graph are present.)	3. Because this is a quadratic function, it could have two x-intercepts. Pick x-values that give a friendly window including the x-intercepts. In this case, using -9.4 to 9.4 gives the graph: 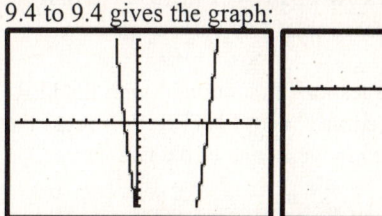
4. To find the point(s) where the graph crosses the x-axis and the equation has solutions, use 2nd CALC, 2 :ZERO . Answer the question "*left bound?*" with ENTER after moving the cursor close to and to the left of an x-intercept. Answer the question "*right bound?*" with ENTER after moving the cursor close to and to the right of this x-intercept. To the question "*guess?*" press ENTER. The coordinates of the x-intercept will be displayed. The x-value is the solution to the original equation.	4. Using TRACE shows a solution near x = 5.4. Use 2nd CALC, 2 (ZERO) .
5. The same steps, with 2nd CALC, 2: Root gives the solution on the TI-82.	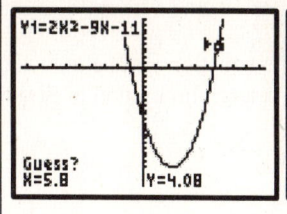 The value of the x-intercept (zero) is 5.5, so a solution is x = 5.5
6. Repeat to get all x-intercepts (and solutions).	6. Repeating the steps above near the second intercept gives the second x-intercept (and solution) as x = -1.

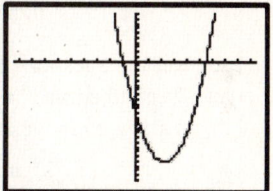

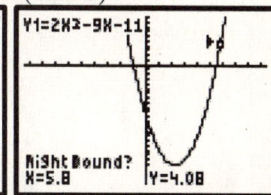

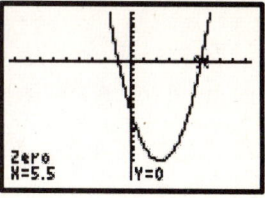

SOLVING AN EQUATION USING THE INTERSECT METHOD	EXAMPLE
To solve an equation (approximately) by the intersect method: 1. Under the Y= menu, assign the left side of the equation to y_1 and the right side of the equation to y_2. 2. Graph the equations using a friendly window that contains the points of intersection of the graphs. Using ZOOM OUT can be used to search for all points of intersection. 3. Press 2nd CALC, 5 (INTERSECT) to find each point of intersection of two curves. Answer the question "first curve?" with ENTER and "second curve?" with ENTER. (Or press the down arrow to move to one of the two curves.) To the question "guess?" move the cursor close to the desired point of intersection and press ENTER. The coordinates of the point of intersection will be displayed. Repeat to get all points of intersection. 4. The solution(s) to the equation will be the values of x from the points of intersection found in Step 3.	Solve $\|2x-1\| = \frac{1}{3}x + 2$ for x. 1. [calculator Y= screen showing Y1=abs(2X-1), Y2=(1/3)X+2] 2. Using ZOOM 4 gives the graphs: [graph screen] 3. [four calculator screens showing First curve? X=0 Y=1; Second curve? X=0 Y=2; Guess? X=-.2 Y=1.9333333; Intersection X=-.4285714 Y=1.8571429] [screen showing Intersection X=1.8 Y=2.6] 4. The solutions to the equation are $x = -.429$ (approximately) and $x = 1.8$.

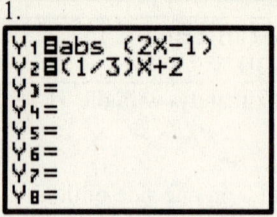

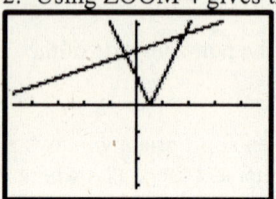

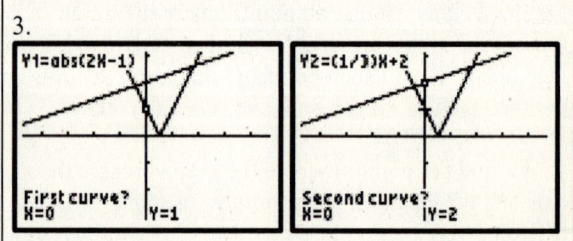

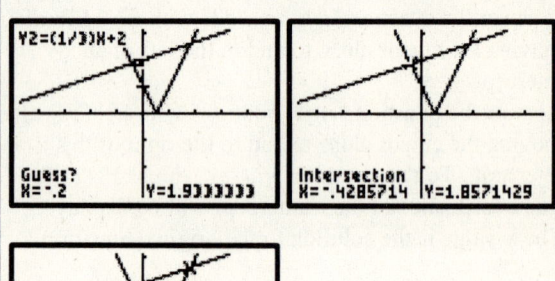

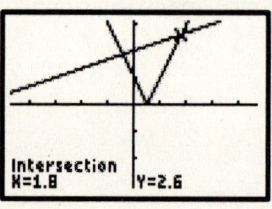

USING SOLVER ON THE TI-83 AND TI-83 PLUS
An equation involving one or more variables can be solved for one variable with the SOLVER function, under the MATH menu of the TI-83.

SOLVING AN EQUATION IN ONE VARIABLE WITH SOLVER ON THE TI-83 AND TI-83 PLUS	EXAMPLE
To solve an equation using SOLVER: 1. Rewrite the equation with 0 on one side. 2. Press MATH 0 (Solver). Press the up arrow revealing EQUATION SOLVER eqn: 0 =, and enter the nonzero side of the equation to be solved. 3. Press the down arrow or ENTER and the variable appears with a value (not the solution). Place the cursor on the variable whose value is sought. Press ALPHA SOLVE (ENTER). The value of the variable changes to the solution of the equation that is closest to that value. 4. To find additional solutions (if they exist), change the value of the variable and press ALPHA SOLVE (ENTER). The value of the variable changes to the solution of the equation that is closest to that value.	Solve $x^2 - 7x = -12$ 1. Write the equation in the form $x^2 - 7x + 12 = 0$ 2. Get the EQUATION SOLVER and enter $x^2 - 7x + 12$. 3. 4.
SOLVING AN EQUATION FOR ONE OF SEVERAL VARIABLES WITH SOLVER 1. Press MATH 0 (Solver). Press the up arrow revealing EQUATION SOLVER eqn: 0 =, and enter the nonzero side of the equation to be solved. 2. Press the down arrow or ENTER and the variables appear. Enter known values for the variables, and place the cursor on the variable whose value is sought. 3. Press ALPHA SOLVE (ENTER). The value of the variable changes to the solution of the equation.	Use $I = PRT$ to find the rate R if an investment of $1000 yields $180 in 3 years. 1. $0 = PRT - I$ 2. The rate is 6%.

VIII. SOLVING SYSTEMS OF EQUATIONS

TRACE can be used to find the intersection of two graphs.

POINTS OF INTERSECTION OF GRAPHS - SOLVING A SYSTEM OF TWO LINEAR EQUATIONS GRAPHICALLY	EXAMPLES
To find the points of intersection of two graphs (or to find the solution of a system of equations graphically). 1. Solve each equation for y and use the "Y=" key with y_1 and y_2 to enter the equations. Graph the equation with a friendly window.	Find the solution graphically: (a) $\begin{cases} 4x + 3y = 11 \\ 2x - 5y = -1 \end{cases}$ 1. $y_1 = 11/3 - (4/3)x$ $y_2 = (2/5)x + 1/5$
2.(a) If the two lines intersect in one point, the coordinates give the x- and y-values of the solution. To find or approximate the intersection, use TRACE with a friendly window. Pressing the up and down arrows moves the cursor from one line to the other. If TRACE gives equal y-values on both lines, this y-value and the x-value is the solution to the system.	2. Using ZOOM 4 and TRACE gives: 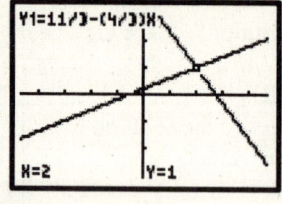 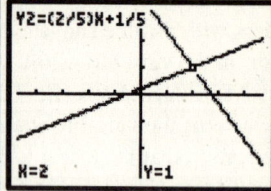 Solution: $x = 2$, $y = 1$
(b) If the two lines are parallel, there is no solution; the system of equations is **inconsistent**. If the lines are parallel, then when solving for y the equations will show that the lines have the same slope and different y-intercepts.	(b) $\begin{cases} 4x + 3y = 4 \\ 8x + 6y = 25 \end{cases}$  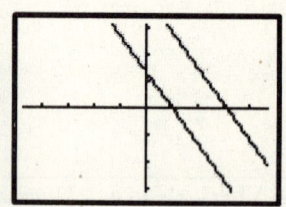 No solution; inconsistent system
(c) If the two graphs of the equations give only one line, every point on the line gives a solution to the system, and the system is **dependent**. The two graphs will be the same graph if, when solving for y to use the graphing calculator, the equations are identical.	(c) $\begin{cases} 2x + 3y = 6 \\ 4x + 6y = 12 \end{cases}$  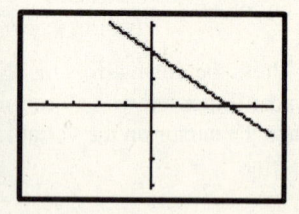 Many solutions; dependent system

The graphing techniques discussed previously can also be used to find the intersection of non-linear curves. There may be more than one solution to systems containing non-linear equations.
Methods other than TRACE can be used to find the intersection of two graphs.

SOLUTION OF SYSTEMS OF EQUATIONS - USING THE INTERSECT METHOD	EXAMPLE
To solve two equations simultaneously using the intersect method:	Solve the system $$\begin{cases} y = 2x^2 - 3x + 2 \\ y = x^2 + 2x + 8 \end{cases}$$
1. Solve each equation for y and use the Y= key with y_1 and y_2 to enter the equations. Graph the equation with a friendly window. Use the graphs to determine how many points of intersection (solutions) there are and approximately where they are.	1. The graphs of the functions are: 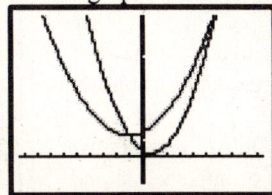
2. Use 2nd CALC, 5 (INTERSECT) to find each point of intersection of two curves. Answer the question "first curve?" with ENTER and "second curve?" with ENTER. To the question "guess?" move the cursor close to the desired point of intersection and press ENTER. The coordinates of the point of intersection will be displayed. Repeat to get all points of intersection.	2. Using 2nd CALC, 5 and answering the questions gives a point of intersection at x = –1, y = 7. 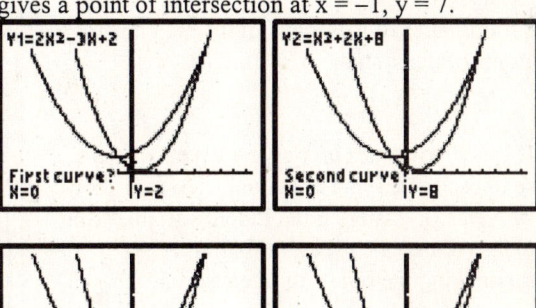 Repeating the process near x = 5 gives the other point of intersection at x = 6, y = 56.
3. The coordinates of each point of intersection give the x- and y-values of the solutions to the system of equations.	3. The solutions to the system of equations are x = 6, y = 56 and x = –1, y = 7.

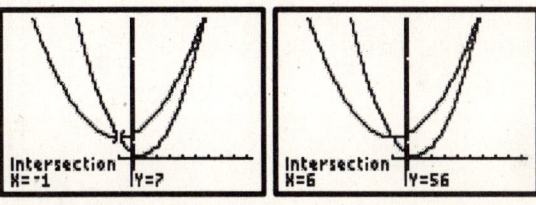

SOLUTION OF SYSTEMS OF EQUATIONS – FINDING OR APPROXIMATING USING TABLE	EXAMPLE
To solve a system of equations using a table: 1. Solve each equation for y and use the Y= key with y_1 and y_2 to enter the equations. Graph the equations. Use the graphs to determine how many points of intersection (solutions) there are and approximately where they are. 2. Use 2nd TABLE SET to build a table that contains values near the x coordinate of a solution. 3. The x-values resulting in EQUAL y-values are the x-coordinates of the points of intersection and the corresponding y-values are the y-coordinates of the intersection points. Use up or down arrows to move the table to all necessary x-values. The coordinates of the points of intersection are the solutions of the system. 4. If specified values of x give y values that are close to each other, but not equal, changing the ΔTbl value or changing the Indpnt variable from Auto to Ask may be useful in finding or approximating the points of intersection.	Solve the system $$\begin{cases} y = 2x^2 - 3x + 2 \\ y = x^2 + 2x + 8 \end{cases}$$ 1. The graph shows two solutions. 2. 3. The solutions are x = -1, y = 7 and x = 6, y = 56. 4. To find the intersection of y = x + 5 and y = 7 - 2x with TABLE SETUP set on Indpnt: Auto, ΔTbl must be set at 1/3. TABLE set with Indpnt: Ask could be used to test values that approach the intersection, until it is found. The solution is x = 0.6667.

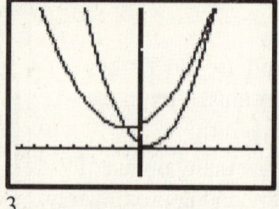

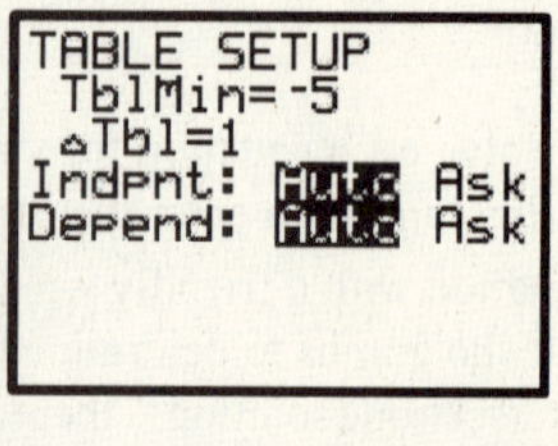

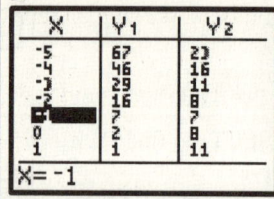

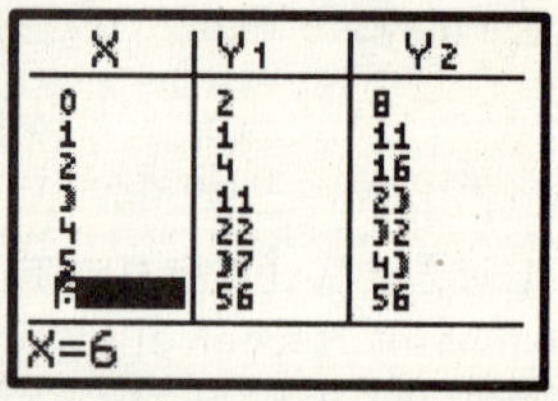

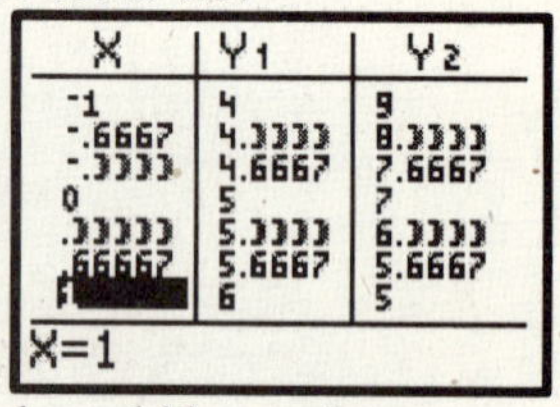

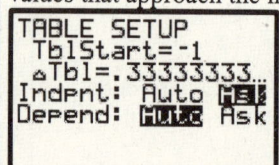

IX. SPECIAL FUNCTIONS; QUADRATIC FUNCTIONS

GRAPHS OF SPECIAL FUNCTIONS	EXAMPLES

1. Linear function: $y = ax + b$
   Graph is a line.
 	
   1. (a) Identity function: $y = x$
      A special linear function.
 	
2. Power Functions $y = ax^b$
 	
(a) Power Functions with $b > 1$
   i. When b is even, the graph is similar to the graph of $f(x) = x^2$.
   ii. When b is odd, the graph is similar to the graph of $f(x) = x^3$.
 	
   The greater the value of n, the flatter the graph is on the interval $[-1, 1]$.
 	
(b) Power Functions $y = ax^b$ with $0 < b < 1$
   (Root functions)
 	
3. Polynomial Functions:
 	
   $y = a_n x^n + a_{n-1} x^{n-1} + ... + a_1 x + a_0$
   All powers of x are positive integers
 	
   Highest power even $\Rightarrow$
      odd number of turns
   Highest power odd $\Rightarrow$
      even number of turns
 	
4. Rational Functions
   Ratio of two polynomials
 	
   4(a). Rectangular hyperbola: $y = \dfrac{1}{x}$
   A special rational function
	1. $y = 4x - 3$ 1(a). $y = x$
	
	2. (a) i. $y = 0.5 x^2$ ii. $y = 0.25 x^3$
	
	2. (b) $y = 4x^{1/2} = 4\sqrt{x}$ $y = 4x^{1/3} = 4\sqrt[3]{x}$
	
	3. $y = 3x^2 + 2x - 5$ $y = 2x^3 + x - 3$
	
	4. $y = \dfrac{x^2}{x-1}$ $y = \dfrac{1}{x}$

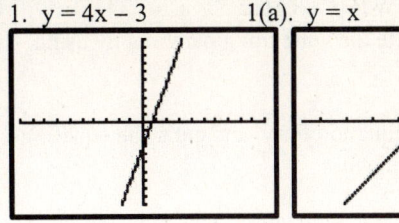

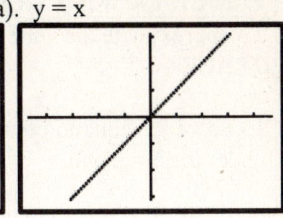

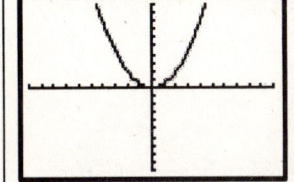

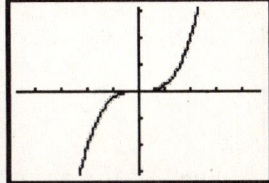

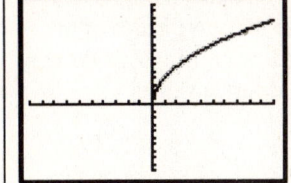

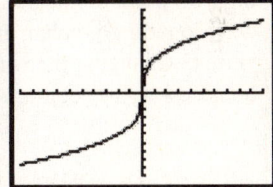

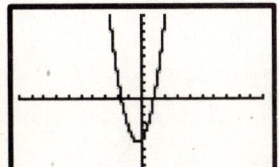

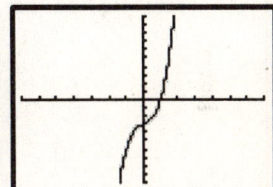

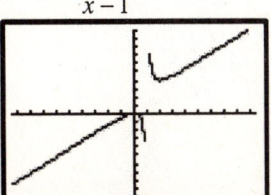

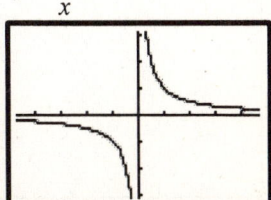

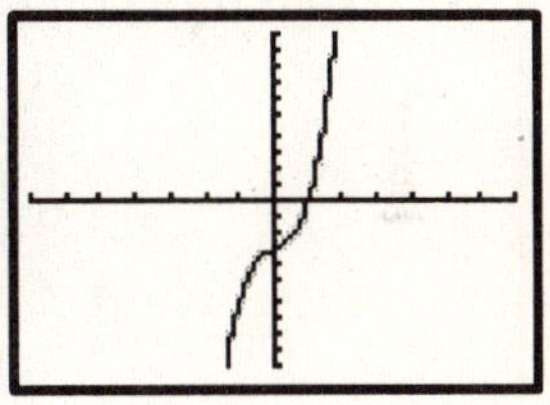

TRACE can be used to approximate the vertex of the graph of a quadratic function.

APPROXIMATING THE VERTEX OF A PARABOLA WITH TRACE	EXAMPLE
To approximate the vertex of a parabola by using TRACE: 1. Solve the equation for y and enter the equation under the Y= menu. 2. Set the window with "friendly" values of x and values of y that are large enough to show the graph is a parabola, and graph the equation. 3. Press TRACE and use the right and left arrows to move the cursor to the vertex (turning point) of the parabola. The x-value that gives the lowest (or highest) point on the parabola, and the corresponding y-value, are the coordinates of the vertex. Changing the window or using ZOOM may be necessary to get more accurate values.	Find the vertex of the graph of $x^2 + 7x - y = 8$. 1. Enter $y_1 = x^2 + 7x - 8$. 2. Using $xmin = -9.4$ and $xmax = 9.4$ and $ymin = -25$ and $ymax = 10$ will show the parabola. 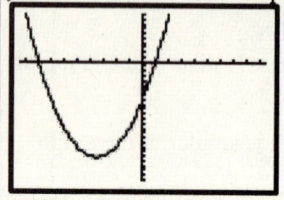 3. TRACE shows the vertex is between -3.4 and -3.6. Changing the window by using ZOOM, Zoom In, or by changing the window to $xmin = -9.4$ and $xmax = 0$, we can trace to the vertex at $x = -3.5$, $y = -20.25$. 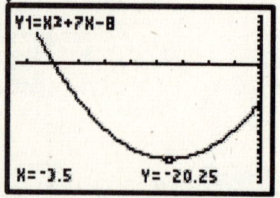
ANOTHER EXAMPLE	Find the vertex of the graph of $-x^2 + 8x - y = 10$. Using $xmin = -9.4$ and $xmax = 9.4$ and $ymin = -10$ and $ymax = 10$ will show the parabola. 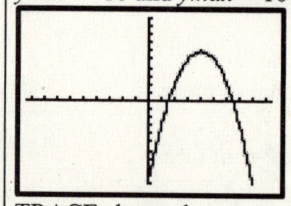 TRACE shows the vertex to be $x = 4$, $y = 6$. 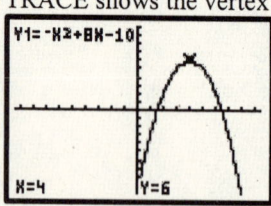

The vertex of a parabola is a maximum or minimum, so we use 2nd CALC and Maximum (or Minimum).

FINDING VERTICES OF PARABOLAS WITH CALC, MAXIMUM OR MINIMUM	EXAMPLE
To find the vertex of a parabola by using CALC, minimum: 1. Solve the equation for y and enter the equation under the Y= menu. 2. Set the window with values of x and values of y that are large enough to show the graph is a parabola, and graph the equation. 3. If the parabola opens up, enter 2nd CALC, 3 (minimum) to locate the vertex. If the parabola opens down, enter 2nd CALC, 4 (maximum). To the question "*left bound?*" ("lower bound" on the TI-82) move the cursor close to and to the left of the vertex and press ENTER To the question "*right bound?*" ("upper bound" on the TI-82) move the cursor close to and to the right of the vertex and press ENTER. To the question "*guess?*" press ENTER. The coordinates of the vertex will be displayed.	Find the vertex of the graph of $x^2 + 7x - y = 8$. 1. Enter $y_1 = x^2 + 7x - 8$. 2. Using a standard window does not show the vertex of the parabola. Changing *ymin* to –25 will show the parabola. 3. The parabola opens up and TRACE shows the vertex is between -3.4 and -3.6. Using 2nd CALC, 3 (minimum), and answering the questions locates the vertex at x = –3.5 y = –20.25. The vertex is at x = –3.5, y = –20.25.

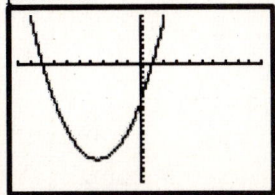

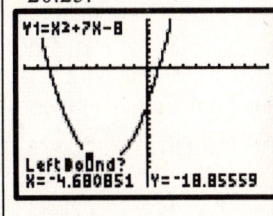

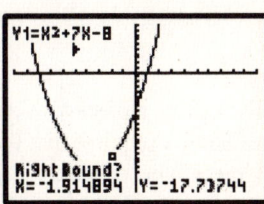

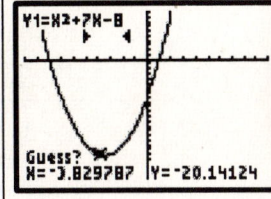

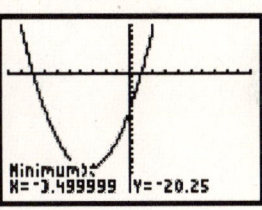

X. PIECEWISE-DEFINED FUNCTIONS

GRAPHING PIECEWISE-DEFINED FUNCTIONS	EXAMPLE
To graph a piecewise-defined function $y = \begin{cases} f(x) \text{ if } x \leq a \\ g(x) \text{ if } x > a \end{cases}$: 1. Under the Y= key, set $y_1 = (f(x))/(x \leq a)$ where $\leq$ is found under 2nd TEST 6. and set $y_2 = (g(x))/(x > a)$ where $>$ is found under 2nd TEST 3. 2. Use an appropriate window, and use ZOOM or GRAPH to graph the function. Evaluating a piecewise-defined function at a given value of x requires that the correct equation ("piece") be selected.	Graph $y = \begin{cases} x+7 \text{ if } x \leq -5 \\ -x+2 \text{ if } x > -5 \end{cases}$ 1. Enter $y_1 = (x + 7)/(x \leq -5)$ and $y_2 = (-x + 2)/(x > -5)$ [calculator screen: Y1■(X+7)/(X≤-5), Y2■(-X+2)/(X>-5)] 2. Use ZOOM 6 to see the graph of the function. [graph screen] To find f(-6), move the cursor on the graph to y_1 and use TRACE, VALUE at -6. [graph screen: Y1=(X+7)/(X≤-5), X=-6, Y=1] To find f(3), move the cursor on the graph to y_2 and use TRACE, VALUE at 3. [graph screen: Y2=(-X+2)/(X>-5), X=3, Y=-1]

40

XI. SCATTERPLOTS AND MODELING DATA

SCATTERPLOTS OF DATA	EXAMPLE
To create a scatterplot of data points: 1. Press STAT and under EDIT press 1:Edit. Enter the x-values in the column headed L1 and the corresponding y-values in the column headed L2. 2. Press 2nd STAT PLOT, 1:Plot 1. Highlight ON, and then highlight the first graph type, Enter Xlist:L1, Ylist:L2, and pick the point plot mark you want. 3. Press GRAPH with an appropriate window or press ZOOM, 9:ZoomStat to plot the data points.	Use AT&T revenues with the number of years past 1980 (x) and revenue in $billions(y) to model the revenue. The data points are (5, 63.1), (6, 69.9), (7, 60.5), (9, 61.1), (10, 62.2), (11, 63.1), (12, 64.9), (13, 67.2) Create a scatterplot of this data. 1. 2. 3.
ADDITIONAL EXAMPLE Once the STAT PLOT has been defined as you want it, it can be turned on and off in the equation editor (Y=) by placing the cursor on Plot1 and pressing ENTER. Remember to turn the Plot1 off when not plotting points, as it may interfere with graphing other equations.	Create a scatterplot for the following data points. The following table gives the average yearly income of male householders with children under 18. Put the number of years past 1960 in column L1 and the corresponding incomes in L2, and create a scatterplot for the data. Year 1969 1972 1975 1978 1981 Income $ 33,749 36,323 33,549 37,575 33,337 Year 1984 1987 1990 1993 1996 Income $ 36,002 34,747 33,769 29,320 31,020

MODELING DATA	EXAMPLE
To find an equation that models data points: 1. Press STAT and under EDIT press 1:Edit. Enter the x-values in the column headed L1 and the corresponding y-values in the column headed L2. 2. Press 2nd STAT PLOT, 1:Plot 1. Highlight ON, and then highlight the first graph Type. Enter Xlist :L1, Ylist:L2, and pick the point plot mark you want. 3. Press GRAPH with an appropriate window or ZOOM, 9:ZoomStat to plot the data points. 4. Observe the point plots to determine what type function would best model the data. 5. Press STAT, move to CALC, and select the function type to be used to model the data. Press the number of this function type. Press ENTER to obtain the equation form and coefficients of the variables. 6. Press the Y= key and place the cursor on y_1. Press the VARS key and press 5:Statistics, then move the cursor to EQ and press 1:RegEQ. The regression equation you have selected will appear as y_1. 7. To see how well the equation models the data, press GRAPH. If the graph does not fit the points well, another function may be used to model the data.	Use AT&T revenues with the number of years past 1980 (x) and revenue in $billions(y) to model the revenue. The data points are (5, 63.1), (6, 69.9), (7, 60.5), (9, 61.1), (10, 62.2), (11, 63.1), (12, 64.9), (13, 67.2) 1. 2. 3. 4. The graph looks like a parabola, so use the quadratic model, with QuadReg. 5. Changing the mode to 3 decimal places and repeating step 5 simplifies the equation. 6. 7.

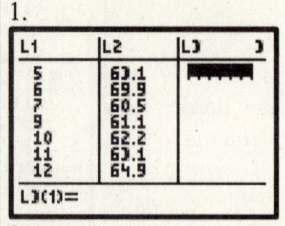

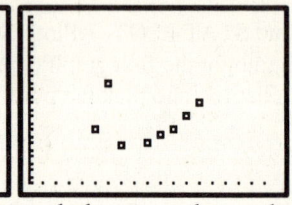

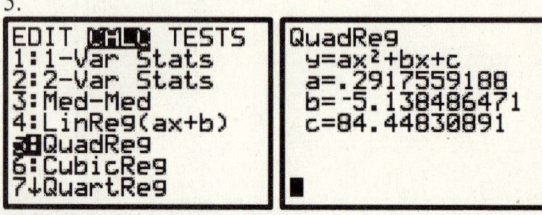

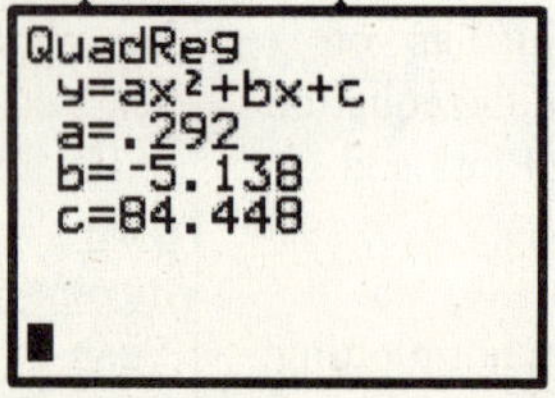

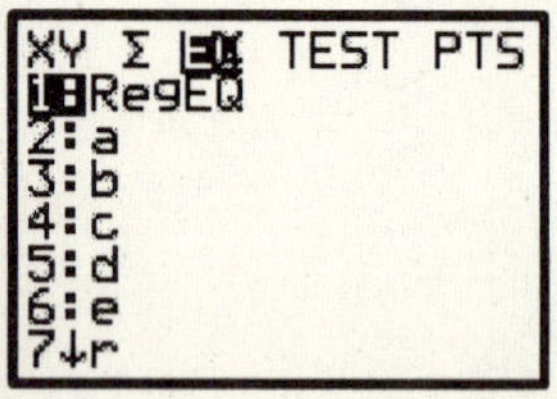

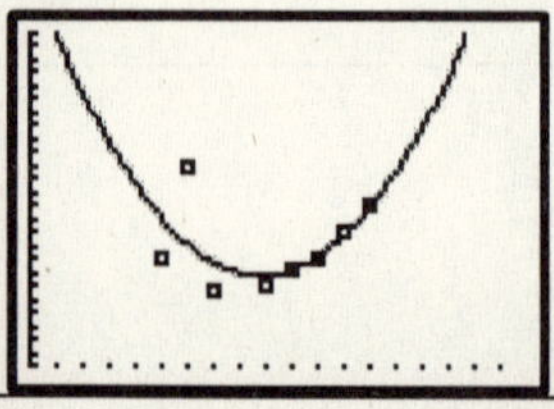

42

XII. MATRICES *

ENTERING DATA INTO MATRICES; THE IDENTITY MATRIX	EXAMPLE
To enter data into matrices:	Enter the matrix below as [A]. $$\begin{bmatrix} 1 & 2 & 3 \\ 2 & -2 & 1 \\ 3 & 1 & -2 \end{bmatrix}$$
1. Press the MATRX key [2nd MATRX on the TI-83 Plus]. 2. Move the cursor to EDIT. Enter the number of the matrix into which the data is to be entered. 3. Enter the dimensions of the matrix, and enter the value for each entry of the matrix. Press ENTER after each entry. 4. To perform operations with the matrix or leave the editor, first press 2nd QUIT. 5. To view the matrix, press MATRX, the number of the matrix, and ENTER.	1., 2. calculator screens showing NAMES MATH EDIT with matrix list 3. Enter 3's to set the dimension, and enter the numbers. MATRIX[A] 3×3 screen 5. [A] screen showing $\begin{bmatrix} 1 & 2 & 3 \\ 2 & -2 & 1 \\ 3 & 1 & -2 \end{bmatrix}$
The n x n matrix consisting of 1's on the main diagonal and 0's elsewhere is called the **identity matrix** of order n and is denoted I_n. To display an identity matrix of order n: 1. Press MATRX, move to MATH, enter 5:identity(, and the order of the identity matrix desired. 2. An identity matrix can also be created by entering the numbers directly with MATRX, EDIT.	Find the identity matrix of order 2. identity(2) $\begin{bmatrix} 1 & 0 \\ 0 & 1 \end{bmatrix}$ Find the identity matrix of order 3. identity(3) $\begin{bmatrix} 1 & 0 & 0 \\ 0 & 1 & 0 \\ 0 & 0 & 1 \end{bmatrix}$

* Note that on the TI-83 Plus, matrices are accessed by pressing 2nd MATRX.

OPERATIONS WITH MATRICES	EXAMPLES
To find the sum of two matrices, [A] and [D]:	Find the sum $\begin{bmatrix} 1 & 2 & 3 \\ 2 & -2 & 1 \\ 3 & 1 & -2 \end{bmatrix} + \begin{bmatrix} 7 & -3 & 2 \\ 4 & -5 & 3 \\ 0 & 2 & 1 \end{bmatrix}$.

1. Enter the values of the elements of [A] using MATRX and EDIT. Press 2nd QUIT. Enter the values of the elements of [D] using MATRX and EDIT. Press 2nd QUIT.

1.
```
MATRIX[A] 3 ×3         MATRIX[D] 3 ×3
[ 1   2   3  ]         [ 7   -3   2 ]
[ 2  -2   1  ]         [ 4   -5   3 ]
[ 3   1  -2  ]         [ 0    2   1 ]
```

2. Use MATRX and NAME, to enter [A] + [D], and press ENTER.

2.
```
[A]+[D]
    [[8  -1   5 ]
     [6  -7   4 ]
     [3   3  -1]]
```

3. If the matrices have the same dimensions, they can be added (or subtracted). If they do not have the same dimensions, an error message will occur.

To find the difference of two matrices, [A] and [D]:

Find the difference.
$\begin{bmatrix} 1 & 2 & 3 \\ 2 & -2 & 1 \\ 3 & 1 & -2 \end{bmatrix} - \begin{bmatrix} 7 & -3 & 2 \\ 4 & -5 & 3 \\ 0 & 2 & 1 \end{bmatrix}$

4. If the matrices are not already entered, enter the values of the elements of [A] and [D] using MATRX and EDIT. After entering the elements forming each matrix, press 2nd QUIT.

4.
```
MATRIX[A] 3 ×3         MATRIX[D] 3 ×3
[ 1   2   3  ]         [ 7   -3   2 ]
[ 2  -2   1  ]         [ 4   -5   3 ]
[ 3   1  -2  ]         [ 0    2   1 ]
```

5. Use MATRIX and NAME, then enter [A] – [D] and press ENTER.

5.
```
[A]-[D]
    [[-6   5   1 ]
     [-2   3  -2 ]
     [ 3  -1  -3]]
```

6. We can multiply a matrix [D] by a real number (scalar) k by pressing
k [D]. (Or k∗[D].)

6. Multiply the matrix [D] by 5.
```
5[D]                   5*[D]
  [[35 -15  10]          [[35 -15  10]
   [20 -25  15]           [20 -25  15]
   [ 0  10   5]]          [ 0  10   5]]
```

44

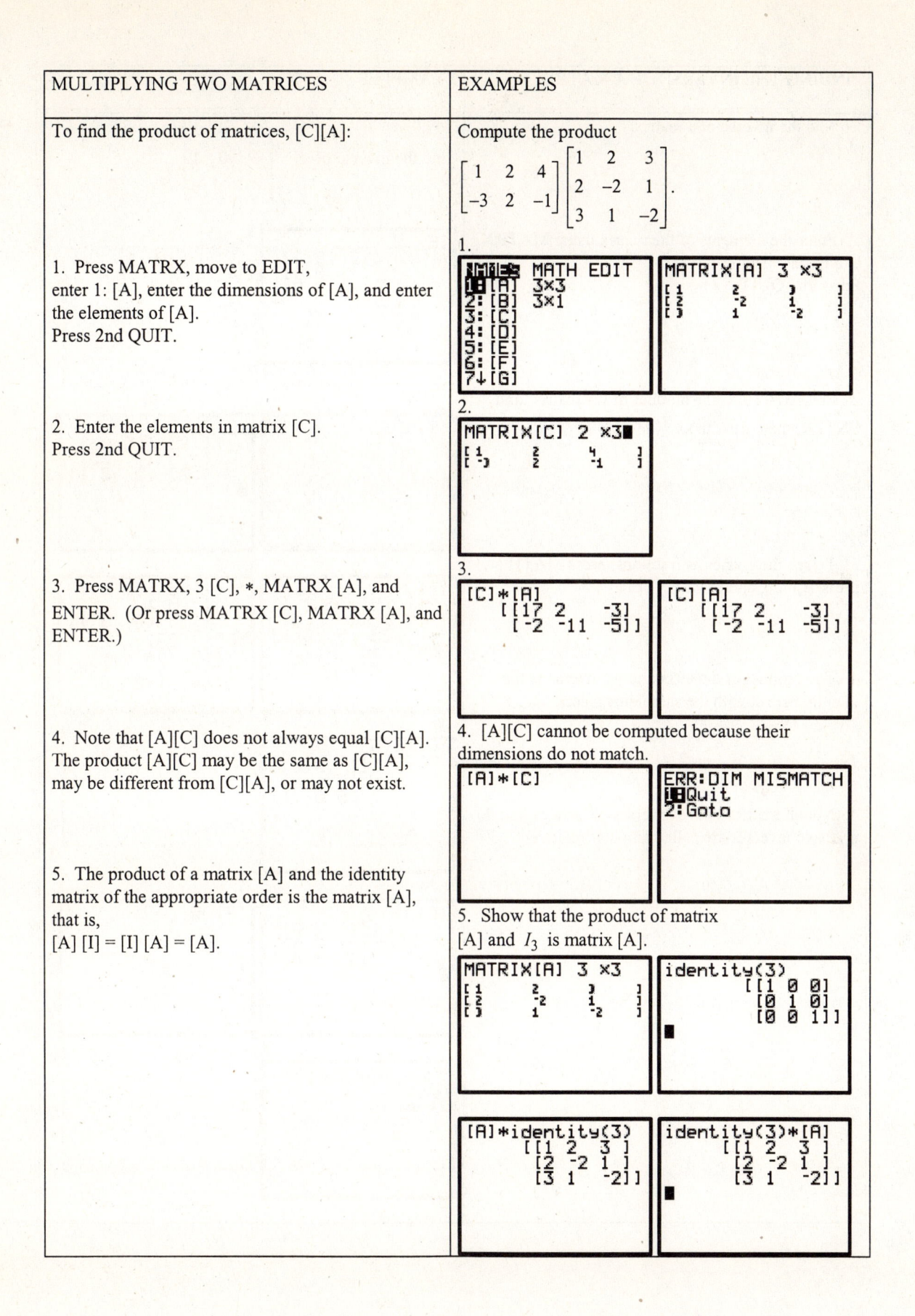

MULTIPLYING TWO MATRICES	EXAMPLES
To find the product of matrices, [C][A]:	Compute the product $\begin{bmatrix} 1 & 2 & 4 \\ -3 & 2 & -1 \end{bmatrix} \begin{bmatrix} 1 & 2 & 3 \\ 2 & -2 & 1 \\ 3 & 1 & -2 \end{bmatrix}$.
1. Press MATRX, move to EDIT, enter 1: [A], enter the dimensions of [A], and enter the elements of [A]. Press 2nd QUIT.	1.
2. Enter the elements in matrix [C]. Press 2nd QUIT.	2.
3. Press MATRX, 3 [C], ∗, MATRX [A], and ENTER. (Or press MATRX [C], MATRX [A], and ENTER.)	3.
4. Note that [A][C] does not always equal [C][A]. The product [A][C] may be the same as [C][A], may be different from [C][A], or may not exist.	4. [A][C] cannot be computed because their dimensions do not match.
5. The product of a matrix [A] and the identity matrix of the appropriate order is the matrix [A], that is, [A] [I] = [I] [A] = [A].	5. Show that the product of matrix [A] and I_3 is matrix [A].

FINDING THE INVERSE OF A MATRIX	EXAMPLES
To find the inverse of a matrix:	Find the inverse of $E = \begin{bmatrix} 2 & 0 & 2 \\ -1 & 0 & 1 \\ 4 & 2 & 0 \end{bmatrix}$.
| | 1. <br> ```
[E]
    [[2  0  2]
     [-1 0  1]
     [4  2  0]]
``` |
| 1. Enter the elements of the matrix using MATRX and EDIT.
 Press 2nd QUIT. | |
| 2. Press MATRX, the number of the matrix, and ENTER, then press the x^{-1} key and ENTER. | 2.
 ```
[E]-1
```   ```
[E]-1
    [[.25 -.5 0]
     [-.5 1 .5]
     [.25 .5 0]]
``` |
| 3. To see the entries as fractions, press MATH and press 1: Frac, and press ENTER. | 3.
 ```
[E]-1
    [[.25 -.5 0]
     [-.5 1 .5]
     [.25 .5 0]]
Ans▶Frac
```   ```
    [[.25 -.5 0]
     [-.5 1 .5]
     [.25 .5 0]]
Ans▶Frac
    [[1/4 -1/2 0 ...
     [-1/2 1   1/2...
     [1/4   1/2 0 ...
``` |
| 4. The product of a matrix and its inverse is the identity matrix with the same dimension. | 4. <br> ```
[E][E]-1
    [[1 0 0]
     [0 1 0]
     [0 0 1]]
``` |
| 4. Not all matrices have inverses. Matrices that do not have inverses are called singular matrices. | 4.
 ```
[F]
    [[1 2 0]
     [-1 2 0]
     [2 3 0]]
```   ```
[F]-1■
``` <br><br> ```
ERR:SINGULAR MAT
1:Quit
2:Goto
``` |

46
```

DETERMINANT OF A MATRIX; TRANSPOSE OF A MATRIX	EXAMPLES
To find the determinant of a matrix [A]:	Find the determinant of the matrix A given below.$$A = \begin{bmatrix} 2 & 3 & 0 & 1 & 3 \\ 1 & 0 & 2 & 3 & 1 \\ 0 & 1 & 0 & 2 & 3 \\ 1 & 0 & 2 & 2 & 1 \\ 1 & 0 & 0 & 0 & 3 \end{bmatrix}$$
1. Press MATRX, move to EDIT, enter 1: [A], enter the dimensions of [A], and enter the elements of [A]. Press 2nd QUIT when all elements are entered. To view matrix A, Press MATRX, 1:[A}.	1.
2. Press MATRX, move to MATH, enter 1: det(, press MATRX, 1:[A], and press ENTER. This gives det([A]). 3. If det([B]) = 0, the matrix is singular (it has no inverse).	2.
To find the transpose of a matrix [A]: 1. Press MATRX, move to EDIT, enter 1: [A], enter the dimensions of [A], and enter the elements of [A]. Press 2nd QUIT when all elements are entered. To view matrix A, Press MATRX, 1:[A}.	Find the transpose of the matrix A above. 1.
2. Press MATRX, press 1:[A], press MATRX, move to MATH, enter 2: ^{T}, and press ENTER. This gives transpose ([A]).	2.

SOLVING SYSTEMS OF LINEAR EQUATIONS WITH UNIQUE SOLUTIONS	EXAMPLES
To solve a system of equations that has a unique solution: 1. Write each equation with the constant on one side and the variables aligned on the other side. Enter the coefficients of the variables into matrix [A], with the coefficients of the x-variables as column 1, the coefficients of the y-variables as column 2, and the coefficients of the z-variables as column 3. This is called the coefficient matrix. 2. Enter the constants into a second matrix [B]. 3. Multiply the inverse of the coefficient matrix times the matrix of constants. The product is the solution to the system. 4. If the solution to the system is not unique or does not exist, an error statement will occur when using this method. 5. A system does not have a unique solution if the inverse of the coefficient matrix does not exist. (Or equivalently, the determinant of the matrix is 0.)	Solve the system $\begin{cases} x + 2y + 3z = 0 \\ 2x - 2y + z = 7 \\ 3x + y - 2z = -1 \end{cases}$. 1. MATRIX[A] 3×3 2. MATRIX[B] 3×1 3. [A]⁻¹*[B] → [A]⁻¹*[B] = [[1],[-2],[1]] The solution is x = 1, y = -2, z = 1. 4. The system below has no solution. $\begin{cases} x + 2y + 3z = 0 \\ 2x - 2y - z = 7 \\ 3x + 2z = -1 \end{cases}$ MATRIX[G] 3×3 [G]⁻¹[B] ERR:SINGULAR MAT 1:Quit 2:Goto 5. NAMES MATH EDIT 1:det(2:ᵀ 3:dim(4:Fill(5:identity(6:randM(7↓augment(det([G]) = 0

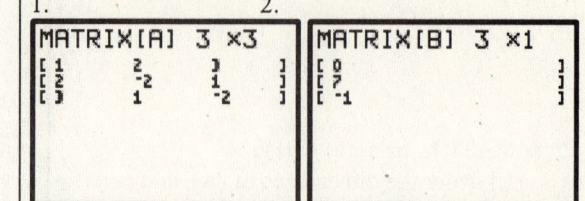

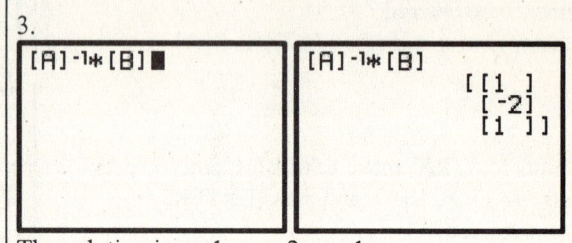

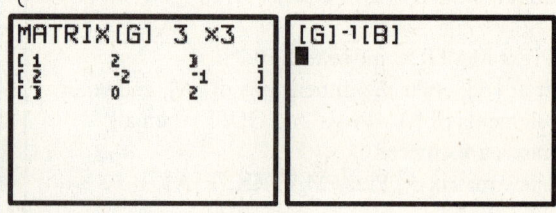

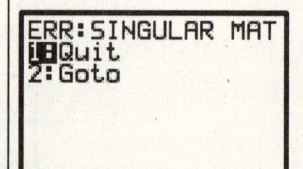

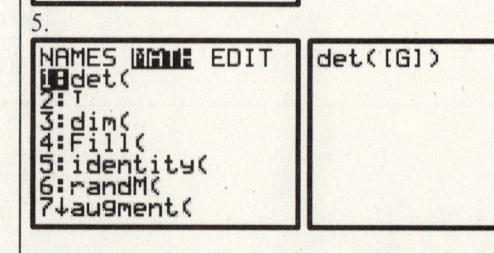

SOLUTION OF SYSTEMS OF 3 LINEAR EQUATIONS IN 3 VARIABLES	EXAMPLE
To solve a system of three equations in three variables:	Solve the system: $\begin{cases} 2x - 2y = 6 \\ x + 2y + 3z = 9 \\ 3x + 3z = 15 \end{cases}$
1. Create an augmented matrix with the coefficient matrix augmented by the constants.	1. [A] [[2 -2 0 6] [1 2 3 9] [3 0 3 15]]
2. Perform operations that make a 1 in row 1, column 1. Operations used to do this include interchanging rows with MATRX, MATH, C:rowswap(, with the first entry the matrix and the next elements the rows to be interchanged and/or multiplying row 1 with MATRX, MATH, E:*row(, with the elements value, matrix, and row.	2. Interchange row 1 and row 2 of matrix A, using MATRX, MATH, C:rowswap([A], 1,2) rowSwap([A],1,2) [[1 2 3 9] [2 -2 0 6] [3 0 3 15]]

3. Use row 1 only to get zeros in the other entries of column 1. The operation used is frequently MATRX, MATH, F: *row+(, with elements value, matrix, first row, and second row. Use 2nd ANS to enter the required matrix.	3. Multiply row 1 of the matrix by -2, add it to row 2, and place the sum into row 2, using MATRX, MATH, F: *row+(-2,ANS,1,2). Repeat this step with -3 and row 3.
	 *row+(-2,Ans,1,2) *row+(-3,Ans,1,3)
	 [[1 2 3 9] [[1 2 3 9]
	  [0 -6 -6 -12] [0 -6 -6 -12]
	  [3 0 3 15]] [0 -6 -6 -12]]

4. Perform operations that make a 1 in row 2, column 2. Use MATRX, MATH, E:*row(, with the elements value, matrix, and row.	4. Use MATRX,MATH E:*row(-1/6,ANS,2) to get a 1 in row 2, column 2. *row(-1/6,Ans,2) [[1 2 3 9] [0 1 1 2] [0 -6 -6 -12]]
5. Use row 2 only to get zeros as the other entries in column 2.	
6. If the bottom row contains all zeros except for the entry in row 3, column 4, there is no solution.	
7. If the bottom row contains all zeros, the system has many solutions. The values for the first two variables are found as functions of the third.	5. Use MATRIX,MATH, F: *row+ *row+(6,Ans,2,3) [[1 2 3 9] [0 1 1 2] [0 0 0 0]]
8. If there is a nonzero element in row 3, use row 3 to solve all equations by substitution.	7. The bottom row contains all zeros, so there are many solutions. From row 1, x + z = 5 or x = 5 − z and from row 2, y + z = 2 or y = 2 − z for any z.

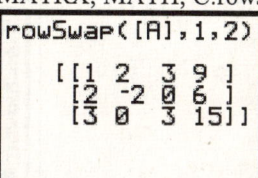

49

SOLUTION OF SYSTEMS - REDUCED ECHELON FORM ON THE TI-83 or TI-83 Plus	EXAMPLE
To solve a system of three equations in three variables by using rref under the MATRX MATH menu:	Solve the system: $\begin{cases} 2x - y + z = 6 \\ x + 2y - 3z = 9 \\ 3x - 3z = 15 \end{cases}$
1. Create an augmented matrix [A] with the coefficient matrix augmented by the constants.	1. 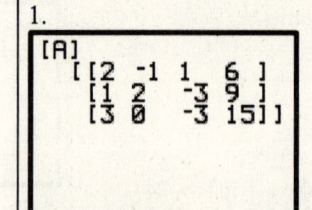 [A] [[2 -1 1 6] [1 2 -3 9] [3 0 -3 15]]
2. Use the MATRX menu to produce a reduced row echelon form of Matrix A, as follows: a. Press MATRX, move to the right to MATH b. Scroll down to B: rref(, and press ENTER, or press ALPHA B. Press MATRX, 1:[A] to get rref([A]). Press ENTER. This gives the reduced echelon form.	2. 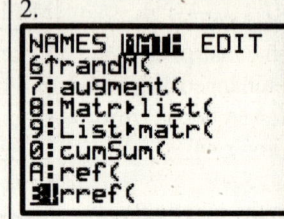 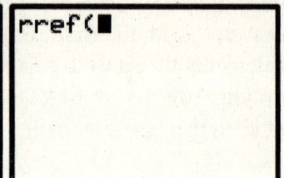 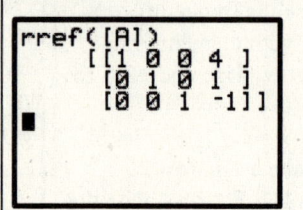 NAMES MATH EDIT rref(6:randM(7:augment(8:Matr▶list(9:List▶matr(0:cumSum(A:ref(B:rref(rref([A]) [[1 0 0 4] [0 1 0 1] [0 0 1 -1]]
3. If each row in the coefficient matrix (first 3 columns) contains a 1 with the other elements 0's, the solution is unique and the number in column 4 of a row is the value of the variable correspnding to a 1 in that row.	3. The solution is unique. x = 4, y = 1, and z = –1
4. If the bottom row contains all zeros, the system has many solutions. The values for the first two variables are found as functions of the third.	
5. If there is a nonzero element in the augment of row 3 and zeros elsewhere in row 3, there is no solution to the system.	

SOLUTION OF SYSTEMS OF LINEAR EQUATIONS: NON-UNIQUE SOLUTIONS	
To solve a system of three equations in three variables on the TI-83 or TI-83 Plus:	Solve the system: $\begin{cases} 2x - 2y = 6 \\ x + 2y + 3z = 9 \\ 3x + 3z = 15 \end{cases}$
1. Create an augmented matrix [A] with the coefficient matrix augmented by the constants.	1. $[A]$ $\begin{bmatrix} 2 & -2 & 0 & 6 \\ 1 & 2 & 3 & 9 \\ 3 & 0 & 3 & 15 \end{bmatrix}$
2. Use the MATRX menu to produce a reduced row echelon form of Matrix A, as follows: a. Press MATRX, move to the right to MATH. b. Scroll down to B: rref(, and press ENTER, or press ALPHA B. Press MATRX, 1:[A] to get rref([A]). Press ENTER. This gives the reduced echelon form.	2. NAMES MATH EDIT ... rref(rref([A]) $\begin{bmatrix} 1 & 0 & 1 & 5 \\ 0 & 1 & 1 & 2 \\ 0 & 0 & 0 & 0 \end{bmatrix}$
3. If each row in the coefficient matrix (first 3 columns) contains a 1 with the other elements 0's, the solution is unique and the number in column 4 of a row is the value of the variable corresponding to a 1 in that row.	3. The bottom row does not contain a 1.
4. If the bottom row contains all zeros, the system has many solutions. The values for the first two variables are found as functions of the third. 5. If there is a nonzero element in the augment of row 3 and zeros elsewhere in row 3, there is no solution to the system.	4. The bottom row contains all zeros, so there are many solutions. From row 1, $x + z = 5$ or $x = 5 - z$ and from row 2, $y + z = 2$ or $y = 2 - z$, for any z.

XIII. SOLVING INEQUALITIES

SOLVING LINEAR INEQUALITIES	EXAMPLE
To solve a linear inequality: 1. Rewrite the inequality with 0 on the right side and simplify. 2. Under the Y= menu, assign the left side of the inequality to y_1, so that $y_1 = f(x)$, where f(x) is the left side. 3. Graph this equation. Set the window so that the point where the graph crosses the x-axis is visible. Note that the graph will cross the axis in at most one point because the graph is of degree 1. (Using ZOOM OUT can help find this point.) 4. Use the ZERO command under the CALC menu to find the x-value where the graph crosses the x-axis. This value can also be found by finding the solution to 0 = f(x) algebraically. 5. Observe the inequality in Step 1. If the inequality is "<", the solution to the original inequality is the interval (bounded by the x-intercept) where the graph is below the x axis. If the inequality is ">", the solution to the original inequality is the interval (bounded by the x-intercept) where the graph is above the x axis. 6. The region above the x-axis and under the graph can be shaded with 2nd DRAW, 7 (Shade), and entering Shade(0, y_1) on the homescreen.	Solve $3x > 6 + 5x$ 1. $3x - 5x - 6 > 0$ $-2x - 6 > 0$ 2. Set $y_1 = -2x - 6$. 3. Using ZOOM 4, the graph is: 4. The x-intercept is x = -3. 5. The inequality is ">", so the solution to the original inequality is the interval (bounded by the x-intercept) where the graph is above the x axis. Thus the solution is x < -3. 6. Using 2nd DRAW, 7 (Shade), and entering Shade(0, y_1) will shade the region above the x-axis and below y_1. The x-interval where the shading occurs is the solution.

SOLVING SYSTEMS OF LINEAR INEQUALITIES	EXAMPLE
To solve a system of inequalities in two variables (approximately):	Solve $\begin{cases} y > |2x-1| \\ y < \frac{1}{3}x + 2 \end{cases}$ for x.
1. Under the Y= menu, assign the right side of the first inequality to y_1 and the right side of the second inequality to y_2. Graph the equations using a friendly window that contains the points of intersection of the graphs.	1. Using ZOOM 4 gives:
2. Visually determine the interval over which the graph of y_2 is above y_1 by using 2nd DRAW, 7 (Shade), and entering Shade(y_1, y_2). Using ZOOM OUT can be used to search for all points of intersection.	2. We seek values of x above $y_1 = |2x-1|$ and below $y_2 = \frac{1}{3}x + 2$.
3. Use 2nd CALC, 5 (intersect) to find one point of intersection of the two curves. Answer the question "first curve?" with ENTER and "second curve?" with ENTER. To the question "guess?" move the cursor close to the desired point of intersection and press ENTER. The coordinates of the point of intersection will be displayed.	3.
4. Repeat to get all points of intersection.	
5. The solution to the system of inequalities will be the interval determined by the values of x from the points of intersection found in Steps 3 and 4.	4.
	5. The solutions to the equation are x = -.429 (approximately) and x = 1.8 The solution is -.429 < x < 1.8 (approx.)

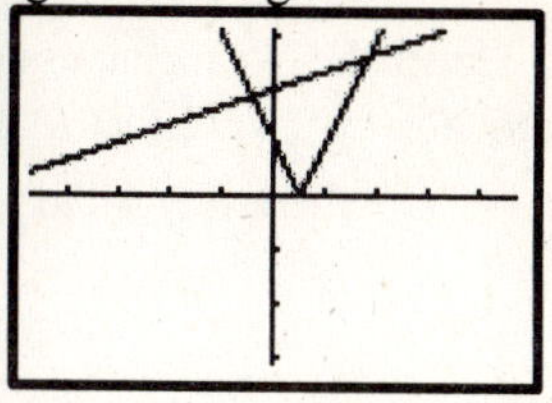

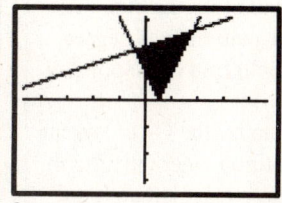

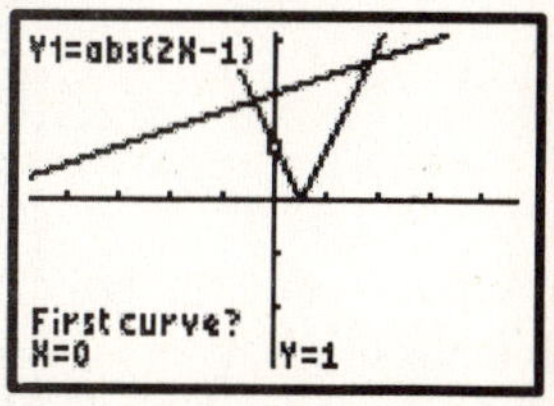

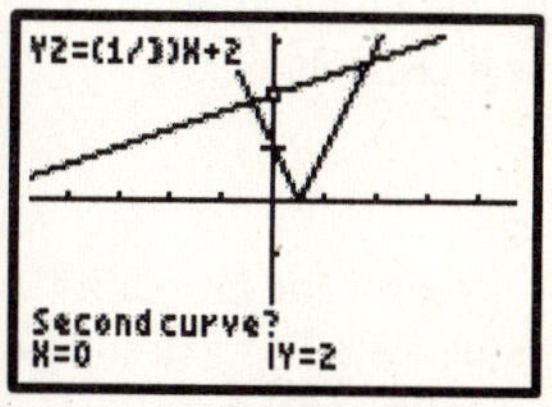

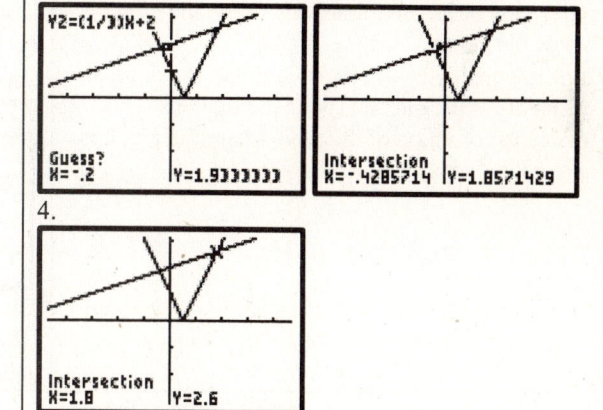

SOLVING QUADRATIC INEQUALITIES	EXAMPLE
To solve a quadratic inequality: 1. Rewrite the inequality with 0 on the right side. 2. Under the Y= menu, assign the left side of the inequality to y_1, so that $y_1 = f(x)$, where f(x) is the left side. 3. Graph this equation. Set the window so that all points where the graph crosses the x-axis are visible. Note that the graph will cross the axis in at most two points because the equation is of degree 2. (Using ZOOM OUT can help find these points.) 4. Use the ZERO command under the CALC menu to find the x-values (one at a time) where the graph crosses the x-axis. These values can also be found by finding the solution to $0 = f(x)$ algebraically. 5. Observe the inequality in Step 1. If the inequality is "<", the solution to the original inequality is the interval (bounded by the x-intercepts) or union of intervals where the graph is below the x-axis. If the inequality is ">", the solution to the original inequality is the interval (bounded by the x-intercepts) or union of intervals where the graph is above the x-axis.	Solve $x^2 - 5x < 6$ 1. $x^2 - 5x - 6 < 0$ 2. 3. Using ZOOM 6, the graph is: 4. The x-intercepts are $x = -1$ and $x = 6$. 5. The graph is below the x-axis between $x = -1$ and $x = 6$. Thus the inequality has solution $-1 < x < 6$. The region below the x-axis can be shaded with 2nd DRAW, 7, (Shade), and entering Shade(y_1, 0) on the homescreen (with ymin = -15).

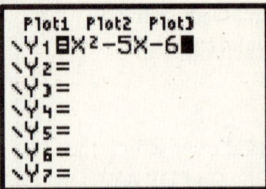

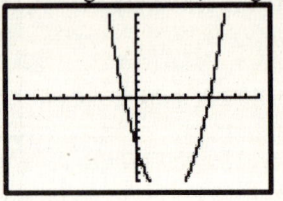

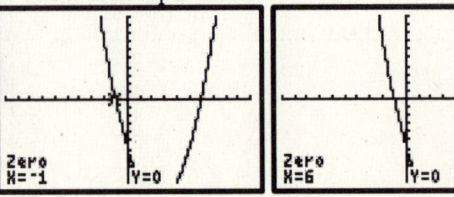

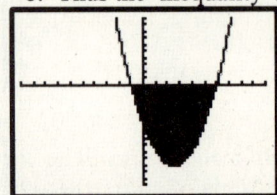

SOLVING QUADRATIC INEQUALITIES - ALTERNATE METHOD	EXAMPLES
To solve a quadratic inequality: 1. Rewrite the inequality with 0 on the right side. 2. Under the Y= menu, assign the left side of the inequality to y_1, so that $y_1 = f(x)$, where $f(x)$ is the left side. 3. Graph this equation. Set the window so that all points where the graph crosses the x-axis are visible. Note that the graph will cross the x-axis in at most two points because the equation is of degree 2. (Using ZOOM, Zoom Out can help find these points.) 4. Use the ZERO command under the CALC menu to find the x-values (one at a time) where the graph crosses the x-axis. These values can also be found by finding the solution to $0 = f(x)$ algebraically. 5. Under the Y= menu, beside $y_2 =$, enter the inequality, with any side containing more than one term enclosed in parentheses. "Turn off" y_1 and graph y_2. The graph displayed resembles a "number line" solution to the inequality, with the x-intercepts as bounds. The zeros found in Step 4 are the boundaries of the inequality.	Solve $x^2 + 3x \geq 10$ 1. $x^2 + 3x - 10 \geq 0$ 2. 3. Using ZOOM 6, the graph is 4. The x-intercepts are x = −5 and x = 2. 5. 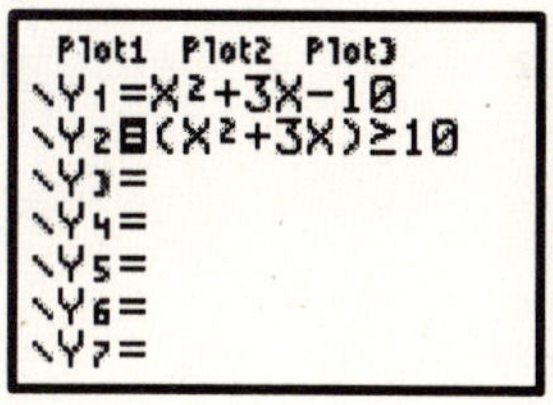 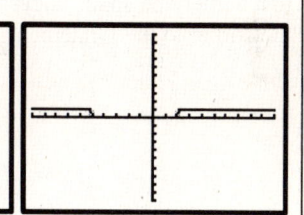 The solution to the inequality $x^2 + 3x \geq 10$ is $x < -5$ or $x > 2$.

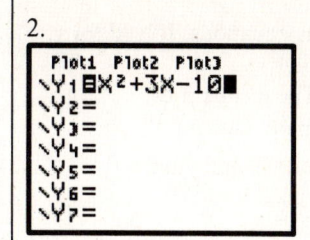

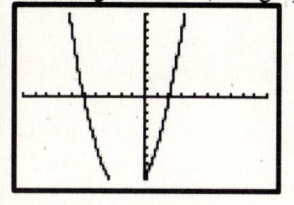

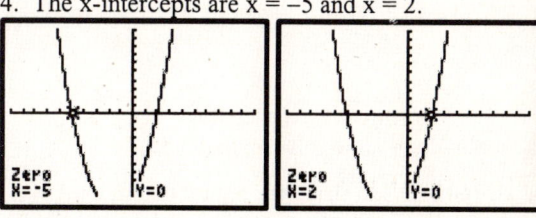

55

XIV. LINEAR PROGRAMMING

GRAPHICAL SOLUTION OF LINEAR PROGRAMMING PROBLEMS	EXAMPLE
To solve a linear programming problem involving two constraints graphically:	Find the region defined by the inequalities $$5x + 2y \leq 54$$ $$2x + 4y \leq 60$$ $$x \geq 0, \ y \geq 0$$
1. Write the inequalities as equations, solved for y.	1. $y = 27 - 5x/2$ $y = 15 - x/2$
2. Graph the equations. The inequalities $x \geq 0, \ y \geq 0$ limit the graph to Quadrant I, so choose a window with xmin = 0 and ymin = 0.	2.
3. Use TRACE or INTERSECT to find the corners of the region, where the borders intersect.	3.
4. Use SHADE to shade the region determined by the inequalities. Shade under the border from x = 0 to a corner and shade under the second border from the corner to the x-intercept.	The corners of the region determined by the inequalities are (0, 15), (6, 12), and (10.8, 0). 4.
5. Evaluating the objective function at the coordinates of each of the corners determines where the objective function is maximized or minimized.	5. At (0, 15), f = 165 At (6, 12), f = 162 At (10.8, 0), f = 54 The maximum value of f is 165 at x = 0, y = 15.

XV. EXPONENTIAL AND LOGARITHMIC FUNCTIONS

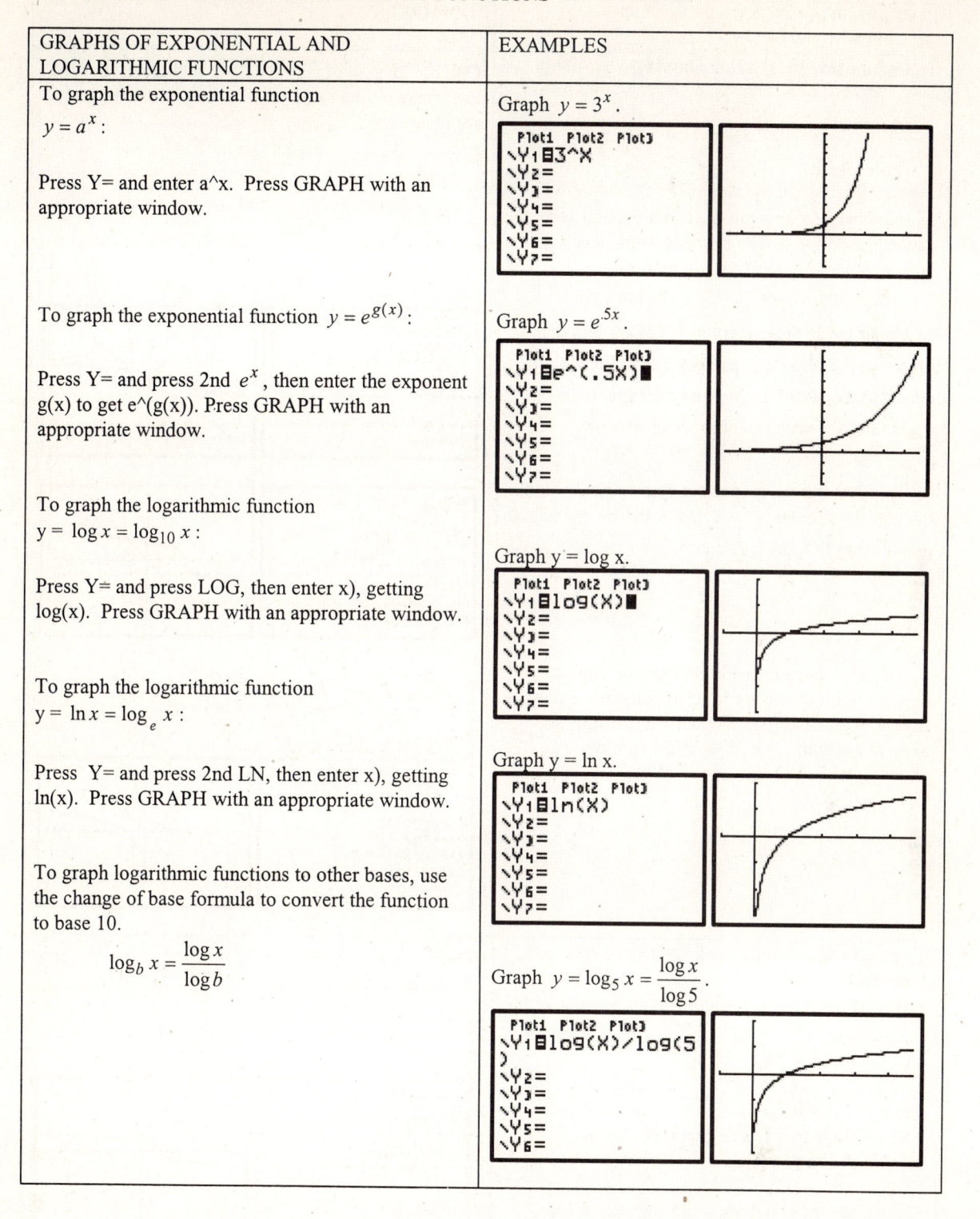

GRAPHS OF EXPONENTIAL AND LOGARITHMIC FUNCTIONS	EXAMPLES
To graph the exponential function $y = a^x$: Press Y= and enter a^x. Press GRAPH with an appropriate window. To graph the exponential function $y = e^{g(x)}$: Press Y= and press 2nd e^x, then enter the exponent g(x) to get e^(g(x)). Press GRAPH with an appropriate window. To graph the logarithmic function $y = \log x = \log_{10} x$: Press Y= and press LOG, then enter x), getting log(x). Press GRAPH with an appropriate window. To graph the logarithmic function $y = \ln x = \log_e x$: Press Y= and press 2nd LN, then enter x), getting ln(x). Press GRAPH with an appropriate window. To graph logarithmic functions to other bases, use the change of base formula to convert the function to base 10. $$\log_b x = \frac{\log x}{\log b}$$	Graph $y = 3^x$. Graph $y = e^{.5x}$. Graph $y = \log x$. Graph $y = \ln x$. Graph $y = \log_5 x = \frac{\log x}{\log 5}$.

INVERSE FUNCTIONS	EXAMPLES
To find the inverse of a function f(x):	Find the inverse of $f(x) = x^3 - 3$ and graph f(x) and its inverse on the same set of axes to show that they are inverses.
1. Write y = f(x).	1. $y = x^3 - 3$
2. Interchange x and y in the equation, and solve the new equation for y. The new equation gives y as the inverse of the original function f(x).	2. Interchanging x and y and solving for y: $x = y^3 - 3 \Rightarrow x + 3 = y^3 \Rightarrow y = \sqrt[3]{x+3}$
3. Under the Y= menu, enter f(x) as y_1 and the inverse function as y_2, and press ENTER with the cursor to the left of y_2 (to make the graph dark). Press GRAPH with an appropriate window.	3.
4. To show that the graphs are symmetrical about the line y = x, enter $y_3 = x$ under the Y= menu and graph using ZOOM,5: Zsquare.	4.
5. To show that a logarithmic function and an exponential function are inverse functions if they have the same base, graph them on the same set of axes, along with y = x. Use a square window.	5. Show that $y = \log_5 x = \dfrac{\log x}{\log 5}$ and $y = 5^x$ are inverse functions.
To graph a function f(x) and its inverse of a function f(x):	Graph the function $f(x) = x^3 - 3$ and its inverse on the same set of axes.
1. Under the Y= menu, enter f(x) as y_1. Choose a square window.	1. Enter $y_1 = x\wedge 3 - 3$
2. Press 2nd DRAW, 8:DrawInv, press VARS, move to Y-VARS, highlight Y-VARS and press ENTER three times. The graph of f(x) and its inverse will be displayed.	2.
3. To clear the graph of the inverse, press 2nd DRAW, 1:ClrDraw.	

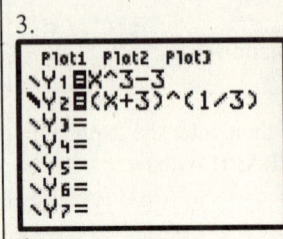

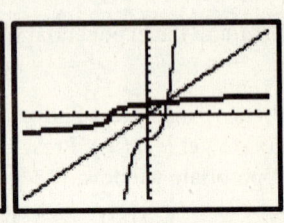

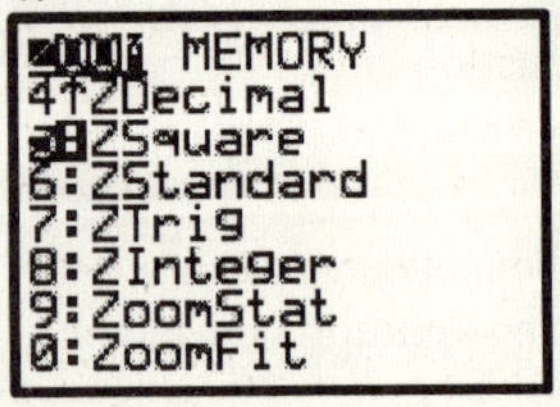

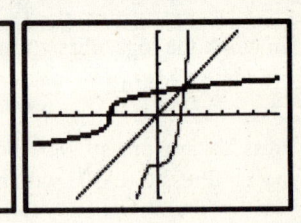

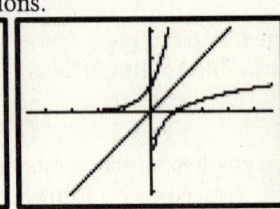

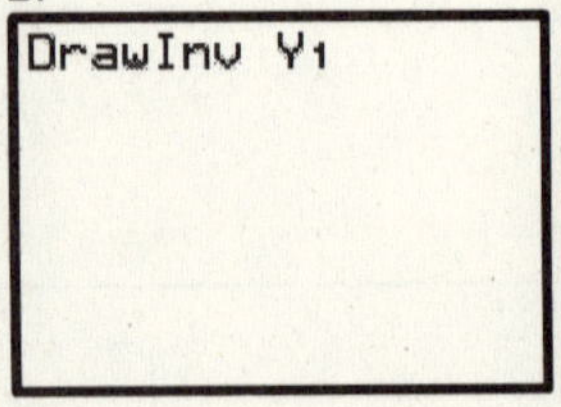

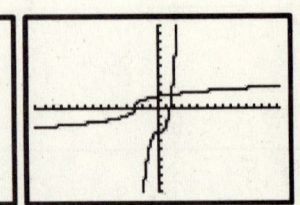

EXPONENTIAL REGRESSION	EXAMPLES
If a scatterplot for data appears to have an exponential shape, the equation that models the data can be created with STAT. To find an equation that models data points: 1. Press STAT and under EDIT press 1:Edit. Enter the x-values in the column headed L1 and the corresponding y-values in the column headed L2. 2. Press 2nd STAT PLOT, 1:Plot 1. Highlight ON, and then highlight the first graph Type. Enter Xlist:L1, Ylist:L2, and pick the point plot mark you want. 3. Press GRAPH with an appropriate window or ZOOM, 9:ZoomStat to plot the data points. 4. Observe the point plots to determine what type function would best model the data. 5. Press STAT, move to CALC, and select the function type to be used to model the data. Press the number of this function type. Press ENTER to obtain the equation form and coefficients of the variables. 6. Press the Y= key and place the cursor on y_1. Press the VARS key and press 5:Statistics, then move the cursor to EQ and press 1:RegEQ. The regression equation you have selected will appear as y_1. 7. To see how well the equation models the data, press GRAPH. If the graph does not fit the points well, another function may be used to model the data.	The following data gives the weekly sales for each of 10 weeks after the end of an advertising campaign, with x representing the number of weeks and y representing sales in thousands of dollars. Find the equation that models this data. x | 1 2 3 4 5 6 7 8 9 10 y | 43 38 33 29 25 22 19 15 14 12 1. 2. 3. 4. The shape of the scatterplot appears to be exponential decay. 5. Use STAT, CALC, 0:ExpReg to get the model. ExpReg y=a*b^x a=50.84239573 b=.86559899 6. 7. Y1=50.842395729436*.86559898996141^X

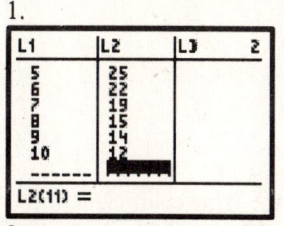

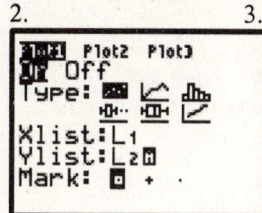

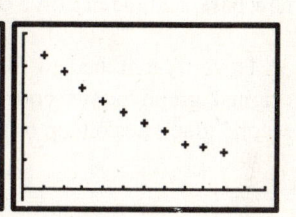

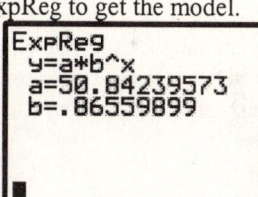

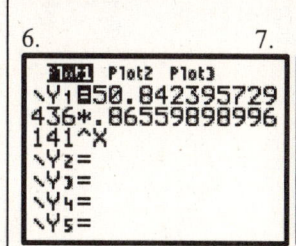

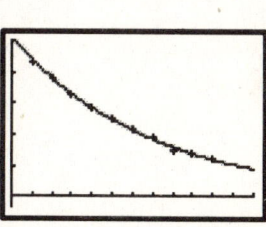

ALTERNATE FORMS OF EXPONENTIAL FUNCTIONS	EXAMPLES
An exponential decay function is frequently written with the base greater than 1 and with a negative exponent. To convert an exponential equation whose base is less than 1 to one whose base is greater than 1:	Convert the equation $y = 50.8424(0.8656)^x$ to an equation with a base greater than 1 and with a negative exponent.
1. Find the reciprocal of the base and change the sign of the exponent, to get a base greater than 1.	1. Enter the base, press the x^{-1} key, and press enter.  .8656⁻¹ 1.155268022
2. Write the equation with the same coefficient, the new base, and the negative of the original exponent.	2. The new form of the equation is $50.8424(1.1553)^{-x}$
3. To verify that the new form is equivalent to the original, graph the two equations and note that the second graph lies on top of the first.	3. 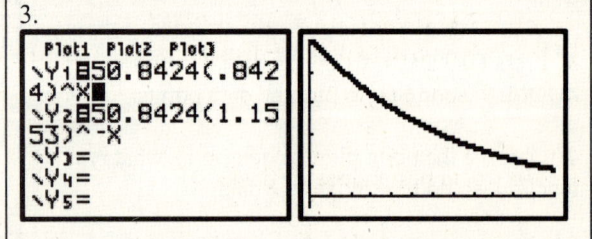
To convert an exponential equation that does not have base e to an exponential equation with base e:	Convert the original equation to an exponential equation with base e:
1. Take the logarithm, base e, of the base.	1.  ln(.8656) -.1443323709
2. The new form of the equation uses this number times the original exponent as the new exponent, has the original coefficient, and has base e.	2. The new exponent is -.1443 times x. The new function is $y = 50.8424e^{-0.1443x}$.
3. To verify that the new form is equivalent, graph both equations on the same set of axes.	3. 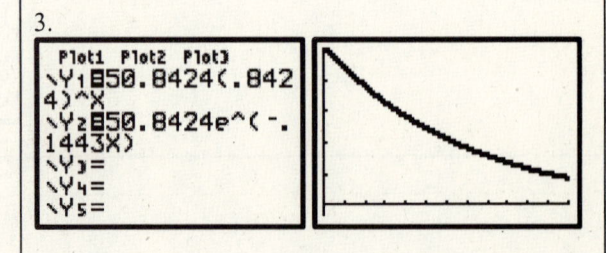

60

LOGARITHMIC REGRESSION	EXAMPLES
If a scatterplot for data appears to have a logarithmic shape, the equation that models the data can be created with STAT. To find an equation that models data points: 1. Press STAT and under EDIT press 1:Edit. Enter the x-values in the column headed L1 and the corresponding y-values in the column headed L2. 2. Press 2nd STAT PLOT, 1:Plot 1. Highlight ON, and then highlight the first graph Type. Enter Xlist:L1, Ylist:L2, and pick the point plot mark you want. 3. Press GRAPH with an appropriate window or ZOOM, 9:ZoomStat to plot the data points. 4. Observe the point plots to determine what type function would best model the data. 5. Press STAT, move to CALC, and select the function type to be used to model the data. Press the number of this function type. Press ENTER to obtain the equation form and coefficients of the variables. 6. Press the Y= key and place the cursor on y_1. Press the VARS key and press 5:Statistics, then move the cursor to EQ and press 1:RegEQ. The regression equation you have selected will appear as y_1. 7. To see how well the equation models the data, press GRAPH. If the graph does not fit the points well, another function may be used to model the data.	The following data gives the millions of hectares (y) destroyed in selected years (x) from 1950. Find the equation that models this data. x | 10 20 30 38 y | 2.21 3.79 4.92 5.77 1. 2. 3. 4. The shape of the scatterplot appears to be logarithmic. 5. Use STAT, CALC, 9:LnReg to get the model. LnReg y=a+blnx a=-3.914351181 b=2.621961591 6. 7.

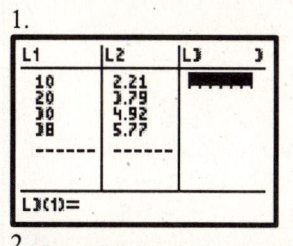

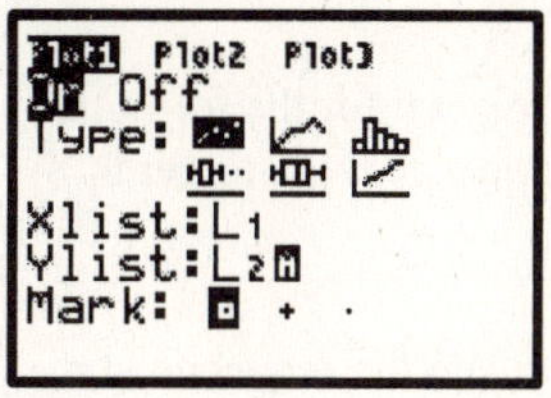

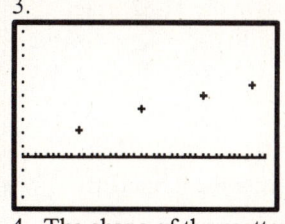

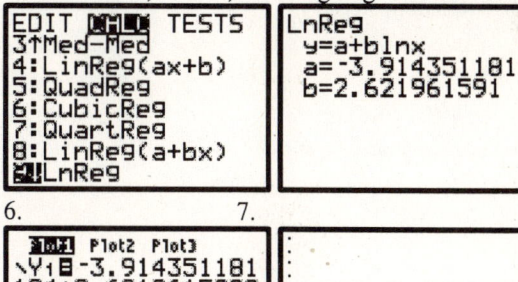

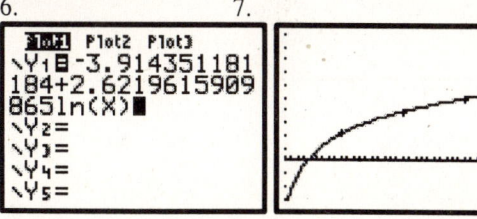

XVI. SEQUENCES

EVALUATING A SEQUENCE	EXAMPLES

To evaluate a sequence for different values of n:
 	
1. Press MODE and highlight Seq. Press ENTER and 2nd QUIT.
 	
2. Store the formula for the sequence (in quotes) in u, using
"formula" STO u.
Press the X,T,θ,n key to get n for the formula, and 2nd 7 to get u.
(On the TI-82, press 2nd n to get n for the formula, and press 2nd VARS, 4: Sequence, 1:u_n to get u_n.)
 	
3. Write u({a,b,c,..}) to evaluate the sequence at a, b, c, ..., and press ENTER. (On the TI-82, enter u_n({a,b,c,..}) and press ENTER.
 	
4. To generate a sequence after the formula is defined, enter
u(nstart, nstop, step), and press ENTER. (On the TI-82, use
u_n(nstart, nstop, step).)
	Evaluate the sequence with nth term $n^2 + 1$ at n = 1, 3, 5, and 9.
	
	1.
	[screen: Normal Sci Eng / Float 0123456789 / Radian Degree / Func Par Pol Seq / Connected Dot / Sequential Simul / Real a+bi re^θi / Full Horiz G-T]
	
	2.
	[screen: "n^2+1"→u]
	
	3.
	[screen: "n^2+1"→u
	              Done
	u({1,3,5,9})
	    {2 10 26 82}]
	
	4. For this sequence, evaluate every third term beginning with the second term and ending with the eleventh term.
	[screen: "n^2+1"→u
	              Done
	u(2,11,3)
	    {5 26 65 122}]

ARITHMETIC SEQUENCES nth TERMS AND SUMS	EXAMPLES
To find the nth term of an arithmetic sequence with first term a and common difference d: 1. Press MODE and highlight Seq. Press ENTER and press 2nd QUIT. 2. Press Y=. At u(n) =, enter the formula for the nth term of an arithmetic sequence, using the x,T,θ,n key to enter *n*. (Use 2nd *n* on the TI-82.) The formula is $$a + (n - 1) * d,$$ where a is the first term and d is the common difference. On the TI-82, u(n) is denoted $u_n(n)$. (Press 2nd VARS, 4: Sequence, 1:u_n.) 3. Press 2nd QUIT. To find the nth term of the sequence, press 2nd u (above 7) followed by the value of n, in parentheses, to get u(n), then press ENTER (On the TI-82, use $u_n(n)$.) 4. Additional terms can be found in the same manner. To find the sum of the first n terms of an arithmetic sequence: 1. Press MODE and highlight Seq,. Press ENTER and press 2nd QUIT. 2. Press Y=. At v(n) =, enter the formula for the sum of the first n terms of an arithmetic sequence, using the x,T,θ,n key to enter *n*. (Use 2nd *n* on the TI-82.) The formula is $(n/2)(a+(a+(n-1)d))$, where a is the first term and d is the common difference. On the TI-82, v(n) is denoted $v_n(n)$. 3. Press 2nd QUIT. To find the sum of the first n terms of the sequence, press 2nd v (above 8) followed by the value of n, in parentheses, to get v(n), then press ENTER. On the TI-82, find $v_n(n)$. 4. Other sums can be found in the same manner.	Find the 12th term of the arithmetic sequence with first term 10 and common difference 5. 1. 2. Substitute 10 for a and 5 for d. 3. The 12th term. 4. The 8th term. u(12) = 65 u(8) = 45 Find the sum of the first 12 terms of the arithmetic sequence with first term 10 and common difference 5. 1. 2. 3. The sum of the first 12 terms. 4. The sum of the first 8 terms. v(12) = 450 v(12) = 450, v(8) = 220

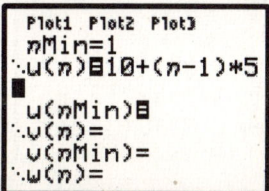

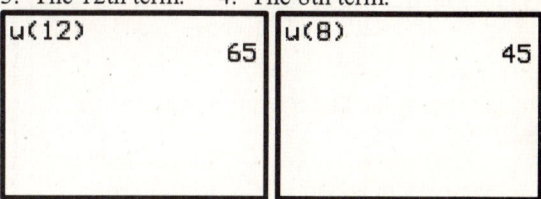

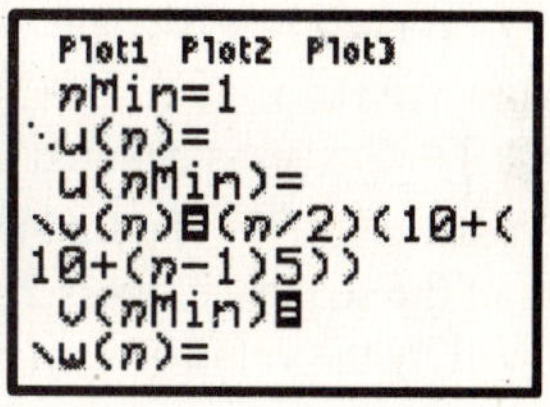

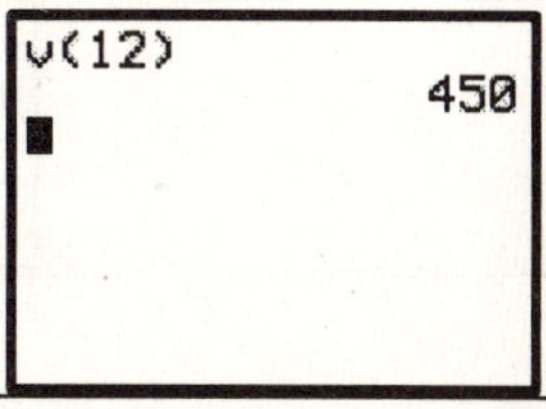

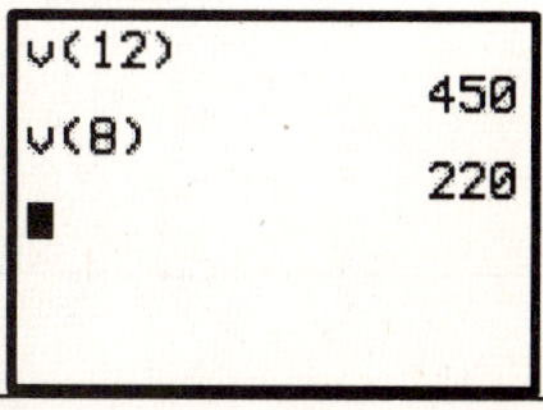

63

GEOMETRIC SEQUENCES nth TERMS AND SUMS	EXAMPLE
To find the nth term of a geometric sequence with first term a and common ratio r: 1. Press MODE and highlight Seq. Press ENTER and press 2nd QUIT. 2. Press Y=. At u(n) =, enter the formula for the nth term of a geometric sequence, using the x,T,θ,n key to enter n. The formula is ar^{n-1}, where a is the first term and r is the common ratio. On the TI-82, u(n) is denoted $u_n(n)$. 3. Press 2nd QUIT. To find the nth term of the sequence, press 2nd u followed by the value of n, in parentheses, to get u(n), then press ENTER (On the TI-82, use $u_n(n)$.) To get a fractional answer, press MATH, 1.: Frac. 4. Additional terms can be found in the same manner. To find the sum of the first n terms of a geometric sequence: 1. Press MODE and highlight Seq. Press ENTER and press 2nd QUIT. 2. Press Y=. At v(n) =, enter the formula for the sum of the first n terms of a geometric sequence, using the x,T,θ,n key to enter n. The formula is $a(1-r^n)/(1-r)$, where a is the first term and r is the common ratio. On the TI-82, v(n) is denoted $v_n(n)$. 3. Press 2nd QUIT. To find the sum of the first n terms of the sequence, press 2nd v (above 8) followed by the value of n, in parentheses, to get v(n), then press ENTER. On the TI-82, find $v_n(n)$. To get a fractional answer, press MATH, 1.: Frac. 4. Other sums can be found in the same manner.	Find the 8th term of the geometric sequence with first term 40 and common ratio 1/2. 1. 2. Substitute 40 for a and (1/2) for r. 3. The 8th term is: 4. The 12 term is: u(8) .3125 u(12) .01953125 Ans▶Frac 5/16 Ans▶Frac 5/256 Find the sum of the first 12 terms of the geometric sequence with first term 40 and common ratio 1/2. 1. 2. Substitute 40 for a and (1/2) for r. 3. The sum of the 4. The sum of the first 12 terms first 8 terms v(12) 79.98046875 v(8) 79.6875 Ans▶Frac 20475/256 Ans▶Frac 1275/16

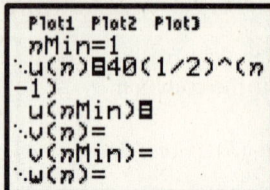

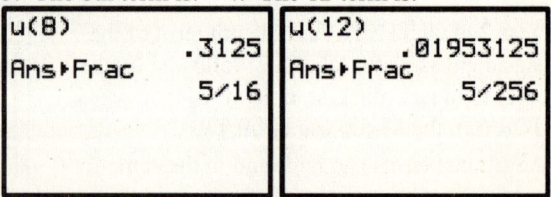

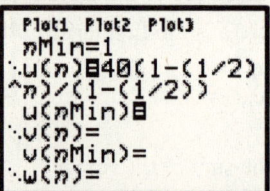

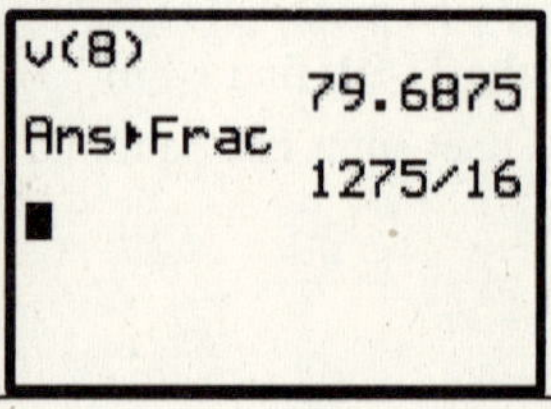

XVII. MATHEMATICS OF FINANCE

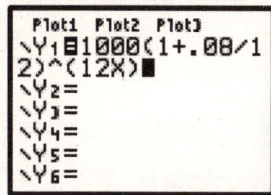

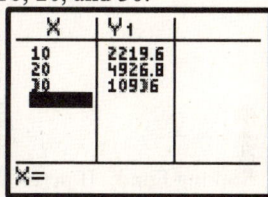

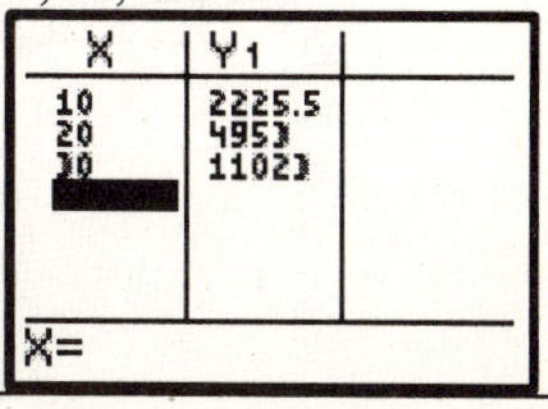

FUTURE VALUE OF AN INVESTMENT	EXAMPLES
| The future values of investments can be found for different rates, times, and compounding periods by entering the formula for S as y₁ in the equation editor, using STO (store) to enter different values for the other variables, and then evaluating y₁.<br><br>1. The future value of an investment of $P invested for t years at a nominal interest rate, r, compounded m times per year, can be found with the formula<br>$$S = P\left(1 + \frac{r}{m}\right)^{mt}.$$<br><br>2. To find the future value of investments for different numbers of years, enter the given values in the formula of step 1, store the formulas as y₁, and read the values in TABLE.<br><br><br><br>3. The future value of an investment of $P invested for t years at a nominal interest rate, r, compounded continuously, can be found with the formula<br>$$S = Pe^{rt}$$ | 1. To find the future value of $1000 invested at 8%, compounded annually and compounded monthly for 10 years, enter the formula as y₁, store the other values, and evaluate y₁. (Use 2nd ENTER to re-enter values for the variables.)<br><br>```
Plot1 Plot2 Plot3      1000→P:.08→R:4→M
\Y1=P(1+R/M)^(M*       :10→T:Y1
T)                            2208.039664
\Y2=                   1000→P:.08→R:12→
\Y3=                   M:10→T:Y1
\Y4=                          2219.640235
\Y5=
\Y6=
```<br><br>2. To find the future value of $1000 invested at 8%, compounded monthly for 10, 20, and 30 years, respectively, enter<br>$$y_1 = 1000\left(1 + \frac{.08}{12}\right)^{12x}$$<br>and use TABLE with x = 10, 20, and 30.<br><br>```
Plot1 Plot2 Plot3    X    Y1
\Y1=1000(1+.08/1     10   2219.6
2)^(12X)             20   4926.8
\Y2=                 30   10936
\Y3=
\Y4=
\Y5=
\Y6=                 X=
```<br><br>3. To find the future value of $1000 invested at 8%, compounded continuously for 10, 20, and 30 years, enter<br>$$y_1 = 1000e^{.08x}$$<br>and use TABLE with x = 10, 20, and 30.<br><br>```
Plot1 Plot2 Plot3    X    Y1
\Y1=1000e^(.08X)     10   2225.5
                     20   4953
\Y2=                 30   11023
\Y3=
\Y4=
\Y5=
\Y6=                 X=
``` |
```

Finance Formulas and TABLE can be used to find future values of annuities for several values of another variable, such as years.

FUTURE VALUES OF ANNUITIES AND PAYMENTS INTO SINKING FUNDS	EXAMPLES
1. Ordinary Annuity: If $R is deposited at the end of each period for n periods in an annuity that earns interest at a rate of i per period, the future value of the annuity is $$S = R\left[\frac{(1+i)^n - 1}{i}\right]$$	1. To find the future value after 10, 20, and 30 years, respectively, of an ordinary annuity with $500 deposited at the end of each quarter at interest rate 8%, compounded quarterly, enter $$y_1 = 500\left[\frac{(1+.02)^{4x} - 1}{.02}\right]$$ and use TABLE with x = 10, 20, and 30.
2. Annuity Due: If $R is deposited at the beginning of each period for n periods in an annuity that earns interest at a rate of i per period, the future value of this annuity due is $$S_{due} = R\left[\frac{(1+i)^n - 1}{i}\right](1+i)$$	2. To find the future value after 10, 20, and 30 years, respectively, of an ordinary annuity with $500 deposited at the beginning of each quarter at interest rate 8%, compounded quarterly, enter $$y_1 = 500\left[\frac{(1+.02)^{4x} - 1}{.02}\right](1+.02)$$ and use TABLE with x = 10, 20, and 30.
3. Sinking Fund: If periodic payments are deposited at the end of each of n periods into an ordinary annuity (or sinking fund) earning interest at a rate of i per period, such that at the end of n periods its value is $S, then the size of each required payment R is $$R = S\left[\frac{i}{(1+i)^n - 1}\right].$$	3. To find the size of deposits made at the end of each month to accumulate money to discharge a debt of $100,000 due in 10, 20, and 30 years, respectively, with interest at 6%, compounded monthly, enter $$y_1 = 100,000\left[\frac{.005}{(1+.005)^{12x} - 1}\right]$$ and use TABLE with x = 10, 20, and 30.

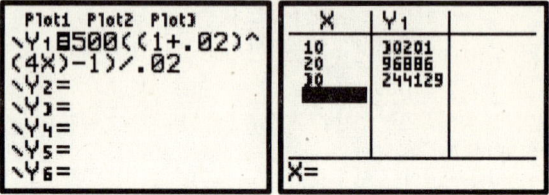

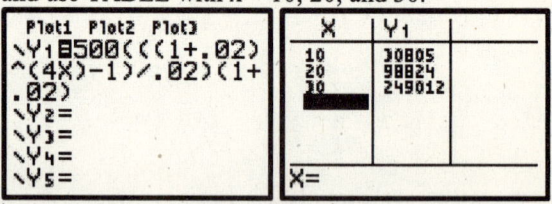

PRESENT VALUE FORMULAS EVALUATING WITH TABLE	EXAMPLES
To find the present value of each of the following types of investments for different numbers of years, enter the given values in the formula, store the formulas in y_1, and read the values in TABLE. 1. If a payment of $R is made at the end of each period for n periods, into (or out of) an annuity that earns interest at a rate of i per period, the present value of the annuity is $$A_n = R\left[\frac{1-(1+i)^{-n}}{i}\right]$$ 2. If a payment of $R is to be made at the beginning of each period for n periods from an account that earns interest rate i per period, the present value of this annuity due is $$A_{due} = R\left[\frac{1-(1+i)^{-n}}{i}\right](i+i)$$ 3. The present value of a deferred annuity of $R per period for n periods deferred for k periods with interest rate i per period is given by $$A_{(n,k)} = R\left[\frac{1-(1+i)^{-n}}{i}\right](1+i)^{-k}$$	1. The present value of an annuity paying $1000 at the end of each month with interest at 12%, compounded monthly, is $$y_1 = 1000\left[\frac{1-(1+.01)^{-12x}}{.01}\right]$$ where x is the number of years. To find the present value for annuities lasting 10, 20, and 30 years, respectively, enter and use TABLE with x = 10, 20, and 30. | X	Y1
10	69701
20	90819
30	97218 | 2. If a payment of $1000 is to be made at the beginning of each period for x years, from an account that earns interest at rate 12%, compounded monthly, the present value of this annuity due is found by entering $$y_1 = 1000\left[\frac{1-(1+.01)^{-12x}}{.01}\right](1+.01).$$ The present value of annuities lasting 10, 20, and 30 years, respectively and using TABLE WITH x = 10, 20, and 30. | X	Y1
10	70398
20	91728
30	98191 | 3. The present value of a deferred annuity of $1000 per month for x years, with interest rate 12%, compounded monthly, after being deferred for 5 years, is $$y_1 = 1000\left[\frac{1-(1+.01)^{-12x}}{.01}\right](1+.01)^{-5\cdot12}$$ To find the present value of annuties lasting 10, 20, and 30 years, respectively and use TABLE with x = 10, 20, and 30. | X	Y1
10	38367
20	49992
30	53514	

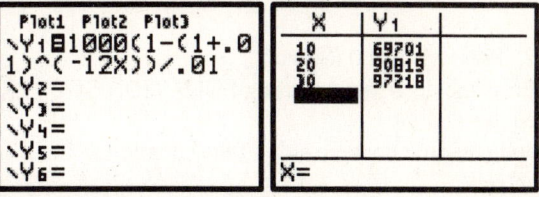

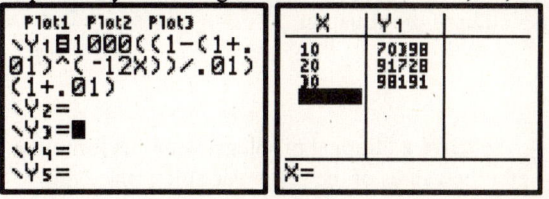

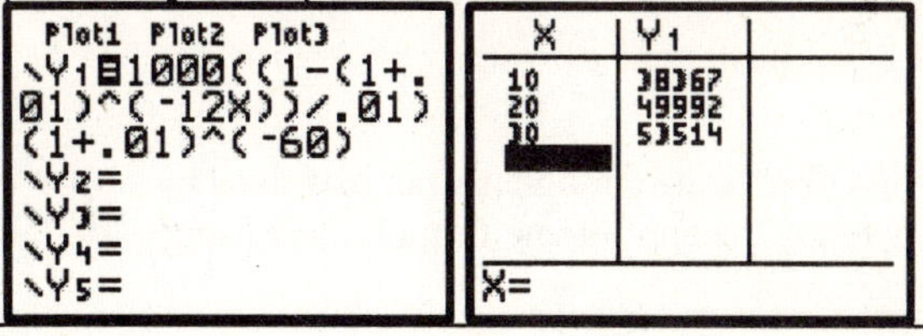

The SOLVER feature under the MATH menu makes it possible to solve the finance formulas for any variable if the values of the other variables are entered.

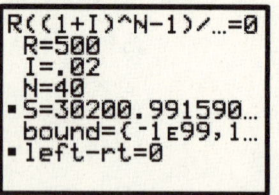

SOLVER AND FINANCE FORMULAS ON THE TI-83 and TI-83 Plus	EXAMPLES
To solve a finance formula for one of the variables:	1. Find the future value after 10 years, of an ordinary annuity with $500 deposited at the end of each quarter at interest rate 8%, compounded quarterly.
1. Rewrite the equation with 0 on one side.	1. Rewrite $S = R\left[\dfrac{(1+i)^n - 1}{i}\right]$ in a form with 0 on one side.
| 2. Press MATH 0 (Solver). Press the up arrow revealing EQUATION SOLVER eq: 0 =, and enter the nonzero side of the equation to be solved. | 2. Press MATH 0 (Solver). Press the up arrow revealing EQUATION SOLVER and enter the equation.<br><br>```
EQUATION SOLVER
eqn:0=R((1+I)^N-
1)/I-S
``` |
| 3. Press the down arrow or ENTER and the variables appear. Enter known values for the variables, place the cursor on the variable whose value is sought, and press ALPHA SOLVE (ENTER). The value of the variable changes to the solution of the equation. | 3. Press the down arrow or ENTER and enter the given values, place the cursor on S and press ALPHA SOLVER.<br><br>```
R((1+I)^N-1)/...=0
 R=500
 I=.02
 N=40
•S=30200.991590…
 bound={-1E99,1…
•left-rt=0
``` |
| 4. To solve additional problems with this formula, enter the values of the given variables, place the cursor on the variable sought, and press ALPHA SOLVE (ENTER). | The future value is $30,200.99.<br><br>4. To find the size of deposits made at the end of each month to accumulate money to discharge a debt of $100,000 due in 10 years, with interest at 6%, compounded monthly, use the same formula, enter the values of the variables, and solve for R.<br><br>```
R((1+I)^N-1)/...=0
•R=610.20501941…
 I=.005
 N=120
 S=100000
 bound={-1E99,1…
•left-rt=0
``` |
| 5. Other finance problems can be solved by entering the appropriate formulas and using SOLVER. | The required deposit is $610.21. |
```

The FINANCE key on the TI-83 and TI-83 Plus can be used to solve many different types of finance problems, including future and present values of annuities and loan payments. The Finance Applications are found under APPS on the TI-83 Plus.

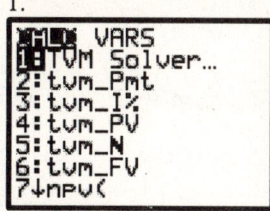

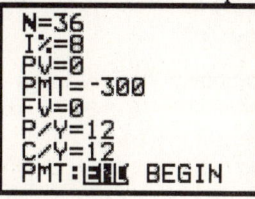

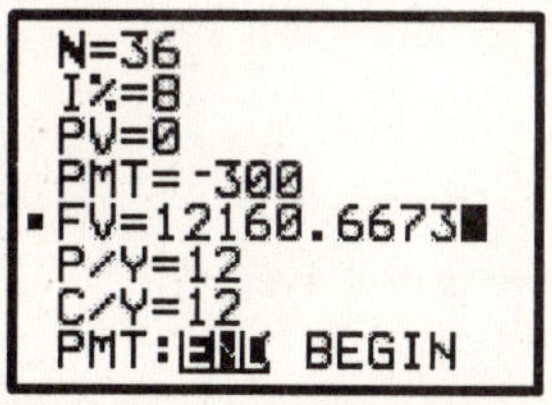

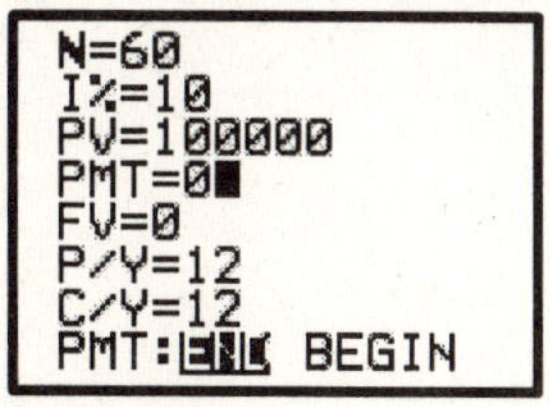

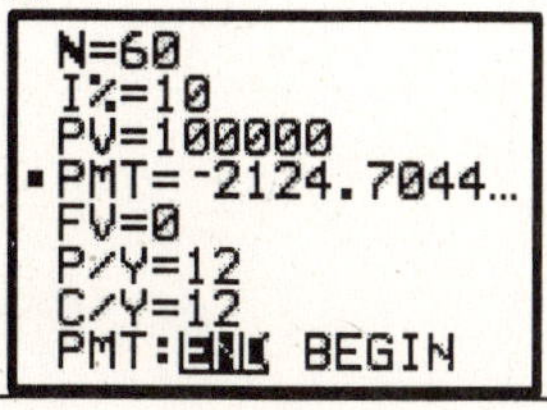

ANNUITIES AND LOANS THE FINANCE KEY ON THE TI-83	EXAMPLES
| ANNUITIES<br><br>To find the future value of an annuity:<br><br>1. Press 2nd FINANCE and from the CALC menu choose 1: TVM Solver.<br><br><br><br>2. Enter the number of periods, N, the interest percent, I%, present value PV, Payment PMT, and the number of compounding periods per year, C/Y.<br><br><br><br><br>3. Place the cursor on FV, the future value, and press ALPHA SOLVE. The future value of the annuity will be displayed.<br><br><br><br>LOANS<br>To find the payment needed to repay a loan:<br>1. Press 2nd FINANCE and from the CALC menu choose 1: TVM Solver.<br>Enter the number of periods, N, the interest percent, I%, the amount of the loan (present value) PV, and the number of payment periods P/Y (the compounding periods per year, C/Y is usually the same). Enter 0 for FV, the future value of the loan (when it is repaid).<br><br>2. Place the cursor on Payment PMT, and press ALPHA SOLVE. The payment of the loan will be displayed.<br>The negative means it is leaving you. | Find the future value of an annuity with a payment of $300 at the end of each of 36 months, with interest at 8%, compounded monthly.<br>1.<br>```
CALC VARS
1:TVM Solver...
2:tvm_Pmt
3:tvm_I%
4:tvm_PV
5:tvm_N
6:tvm_FV
7↓npv(
```<br>2. Enter N = 36, I% = 8, PV = 0, Payment PMT = -300 (negative because it is leaving you), and P/Y and C/Y = 12 for compounding monthly.<br>```
N=36
I%=8
PV=0
PMT=-300
FV=0
P/Y=12
C/Y=12
PMT:END BEGIN
```<br>3.<br>```
N=36
I%=8
PV=0
PMT=-300
•FV=12160.6673
P/Y=12
C/Y=12
PMT:END BEGIN
```<br>Find the monthly payment needed to amortize a loan of $100,000 in 60 months if interest is 10% per year on the unpaid balance.<br>1.<br>```
N=60
I%=10
PV=100000
PMT=0
FV=0
P/Y=12
C/Y=12
PMT:END BEGIN
```<br>2.<br>```
N=60
I%=10
PV=100000
•PMT=-2124.7044...
FV=0
P/Y=12
C/Y=12
PMT:END BEGIN
``` |
```

XVIII. COUNTING AND PROBABILITY

PERMUTATIONS AND COMBINATIONS	EXAMPLES
To compute the number of permutations of n things taken r at a time: 1. Enter n on the home screen. 2. Press MATH, move to PRB, and select 2: nPr. 3. Press ENTER, and enter r. 4. Press ENTER. The answer is displayed. To compute the number of combinations of n things taken r at a time: 1. Enter n on the home screen. 2. Press MATH, move to PRB, and select 3: nCr. 3. Press ENTER, and enter r. 4. Press ENTER. The answer is displayed.	Find the number of permutations of 8 things taken 4 at a time. 1. `8` 2. MATH NUM CPX PRB menu showing 1:rand, 2:nPr, 3:nCr, 4:!, 5:randInt(, 6:randNorm(, 7:randBin(3. `8 nPr 4` 4. `8 nPr 4` → `1680` Find the number of combinations of 8 things taken 4 at a time. 1. `8` 2. MATH NUM CPX PRB menu showing 1:rand, 2:nPr, 3:nCr, 4:!, 5:randInt(, 6:randNorm(, 7:randBin(3. `8 nCr 4` 4. `8 nCr 4` → `70`
ADDITIONAL EXAMPLE	Compare $_{10}P_3$ and $_{10}C_3$. `10 nPr 3` → `720` `10 nCr 3` → `120` $_{10}P_3$ is 6 times larger than $_{10}C_3$.

PROBABILITY USING PERMUTATIONS AND COMBINATIONS	EXAMPLES
To solve a probability problem that involves permutations or combinations:	If a box contains 10 computer chips, of which 5 are defective, what is the probability that two chips drawn from the box are both defective?
1. Analyze the probability problem. It usually involves counting the number of ways an event can occur divided by the total number of possible outcomes.	1. The probability that both are defective is given by the number of ways 2 chips can be drawn from the 5 that are defective divided by the number of ways 2 can be drawn from the 10 in the box.
2. Determine whether permutations or combinations should be used to count the number of ways an event can occur and the total number of possible outcomes. (If order is not important, use combinations.)	2. Use the ratio of combinations because the order in which the chips are drawn is not important. This gives the probability as $$\frac{_5C_2}{_{10}C_2}.$$
3. Enter the ratios of permutations or combinations to find the probability that the event will occur. Using MATH ▸Frac gives the probability as a fraction.	3. ``` 5 nCr 2/10 nCr 2 ■ ``` ``` 5 nCr 2/10 nCr 2 .2222222222 Ans▸Frac 2/9 ```
ADDITIONAL EXAMPLE	If a die is rolled four times, what is the probability that a 5 will occur three times? This is a binomial probability model, solved using n = 5 trials and with the probability of success on each trial p = 1/6. The probability of 3 successes in 4 trials is found using $$_4C_3\left(\frac{1}{6}\right)^3\left(\frac{5}{6}\right).$$ This is evaluated as follows. ``` 4 nCr 3(1/6)³(5/ 6) .0154320988 Ans▸Frac 5/324 ■ ```

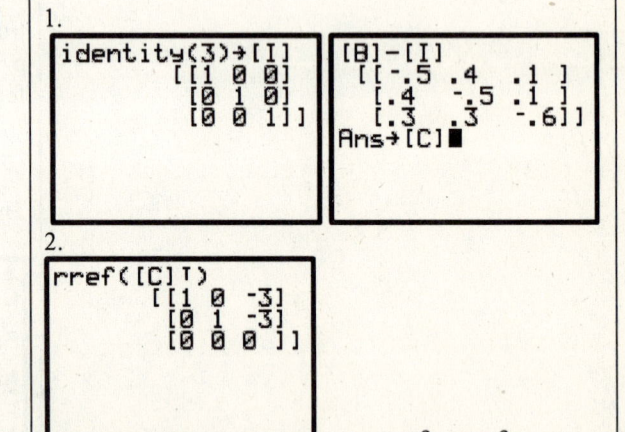

EVALUATING MARKOV CHAINS FINDING STEADY-STATE VECTORS	EXAMPLES
To evaluate a Markov chain:	If the initial-probability vector and the transition matrix for a Markov chain problem are $$[A] = [.4 \quad .4 \quad .2] \text{ and } [B] = \begin{bmatrix} .5 & .4 & .1 \\ .4 & .5 & .1 \\ .3 & .3 & .4 \end{bmatrix},$$ find the probabilities for the fourth state of the chain.
| 1. Enter the initial-probability vector as matrix A and the transition matrix as matrix B. | 1.<br>```
[A]
      [[.4 .4 .2]]
[B]
      [[.5 .4 .1]
       [.4 .5 .1]
       [.3 .3 .4]]
■
``` |
| 2. To find the probabilities for the (n+1)st state, calculate $[A][B]^n$. | 2. The fourth state is $[A][B]^3$.<br>```
[A][B]^3
….428 .428 .144…
``` |
| If the transition matrix contains only positive entries, the probabilities will approach a steady-state vector, which is found as follows: | Find the steady state vector for the Markov chain problem above. |
| 1. Calculate and store $[C] = [B] - [I]$, where [B] is the regular transition matrix and [I] is the appropriately sized identity matrix. | 1.
```
identity(3)→[I]
      [[1 0 0]
       [0 1 0]
       [0 0 1]]
``` ```
[B]-[I]
      [[-.5 .4 .1]
       [.4 -.5 .1]
       [.3 .3 -.6]]
Ans→[C]■
``` |
| 2. On the TI-83, solve $[C]^T = [0]$ by using MATRX, MATH, B:rref($[C]^T$).<br>(Find $[C]^T$ using MATRX, MATH, 2: $^T$.)<br>   On the TI-82 use row operations to solve the equation, because $[C]^T$ is a singular matrix (it has no inverse.) | 2.<br>```
rref([C]T)
      [[1 0 -3]
       [0 1 -3]
       [0 0 0]]
```<br>$\Rightarrow x = 3z, y = 3z$ |
| 3. Choose the solutions that add to 1, because they are probabilities. | 3. $3z + 3z + z = 1$ gives $z = 1/7$, and the probabilities $\begin{bmatrix} \dfrac{3}{7} & \dfrac{3}{7} & \dfrac{1}{7} \end{bmatrix}$ |
```

XIX. STATISTICS

HISTOGRAMS	EXAMPLES

To find a frequency histogram, or more simply, a histogram, for a set of data:
 	
1. Press STAT, EDIT, 1:edit to enter each number in the column headed by L1 and the corresponding frequency of each number in L2.
 	
2. Press 2nd STAT PLOT, 1 (Plot 1). Highlight ON, and then press ENTER with the cursor on the histogram icon.
Enter xlist:L1, Freq:L2.
 	
3. Press ZOOM, 9: ZoomStat or press GRAPH with an appropriate window.
 	
4. If the data is given in interval form, a histogram can be created by using the steps above, with class marks used to represent the intervals.
	Find the frequency histogram for the following scores: 38, 37, 36, 40, 35, 40, 38, 37, 36, 37, 39, 38.
	
	1. Each number can be entered individually, with a frequency of 1, or a frequency table can be used to create the histogram.
	
	4. Create a histogram for the interval data below.
	Interval Frequency
	  1 - 5 0
	  6 - 10 2
	 11 - 15 5
	 16 - 20 1
	 21 - 25 3
	Creating a table with class marks and then using ZOOMSTAT gives the histogram.
	Interval Class Marks Frequency
	  1 - 5 3 0
	  6 - 10 8 2
	 11 - 15 13 5
	 16 - 20 18 1
	 21 - 25 23 3

DESCRIPTIVE STATISTICS	EXAMPLES

To find descriptive statistics for a set of data:	Find descriptive statistics for the following salary data.
	
	    Salary Number Earning
	    $59,000 1
	    30,000 2
	    26,000 7
	    34,000 2
	    31,000 1
	    75,000 1
	    35,000 1

1. Press the STAT key and 1:Edit under EDIT. To clear any elements from a list, place cursor at top of the list and press CLEAR and ENTER. To enter data in a list, enter each number and press ENTER. 2. To find the mean and standard deviation of the data in list L1, press STAT, move to CALC, and press 1: 1-Var Stats, then ENTER.	1. 2. 1-Var Stats $\bar{x}=41428.57143$ $\Sigma x=290000$ $\Sigma x^2=1.4024\text{E}10$ $Sx=18301.70432$ $\sigma x=16944.08572$ $\downarrow n=7$
3. To arrange the data in L1 in descending order, use STAT, EDIT, 3: SortD(L1). Press STAT, EDIT 1:Edit to view the data in descending order.	3. SortD(L1) Done
4. If L2 contains the frequencies of the data in L1, and the data in L1 is in ascending or descending order, the median and mode can easily be read.	4.
5. If L2 contains the frequencies of the data in L1, the mean and standard deviation is found using STAT, CALC 1:1-Vars Stats L1,L2, ENTER.	5. $\bar{x}=34000$ $\Sigma x=510000$ $\Sigma x^2=2.0136\text{E}10$ $Sx=14132.03049$ $\sigma x=13652.83853$ $\downarrow n=15$ ◘1-Var Stats

Probability values of a random variable in probability distributions can be evaluated with the TI-83 calculator.

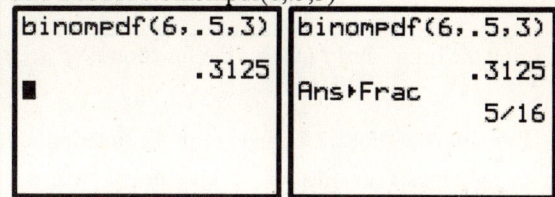

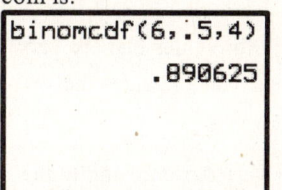

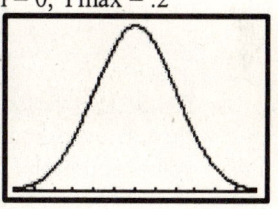

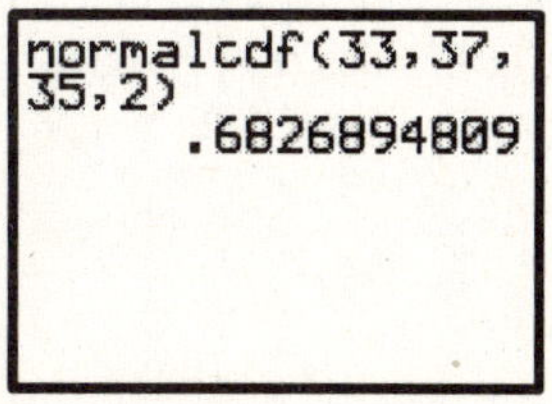

PROBABILITY DISTRIBUTIONS WITH THE TI-83 and TI-83 Plus	EXAMPLES
BINOMIAL DISTRIBUTION 2nd DISTR 0:binompdf(n,p,x) computes the probability at x for the binomial distribution with number of trials n and probability of success p. Using MATH 1:▸Frac gives the probabilities as fractions.	The probability of 3 heads in 6 tosses of a coin is found using 2nd DISTR 0:binompdf(6,.5,3) ```
binompdf(6,.5,3)
.3125
``` ```
binompdf(6,.5,3)
.3125
Ans▸Frac
5/16
``
The probabilities can be computed for more than one number in one command, using 2nd DISTR 0:binompdf(n,p,{$x_1, x_2,..$}). Using MATH 1:▸Frac gives the probabilities as fractions.	The probabilities of 4, 5, or 6 heads in 6 tosses of a coin are: ```
binompdf(6,.5,{4
,5,6})
{.234375 .09375…
``` ```
binompdf(6,.5,{4
,5,6})
{.234375 .09375…
Ans▸Frac
{15/64 3/32 1/6…
``
| 2nd DISTR 0:binomcdf(n,p,x) computes the probability that the number of successes is less than or equal to x for the binomial distribution with number of trials n and probability of success p. | The probability of 4 or fewer heads in 6 tosses of a coin is:<br><br>```
binomcdf(6,.5,4)
           .890625
``` |
| **NORMAL DISTRIBUTION**<br><br>To graph the normal distribution, press Y= and store 2nd DISTR 1:normalpdf(x, $\mu, \sigma$) into $y_1$. Then set the window variables Xmin and Xmax so that the mean $\mu$ falls between them. Press ZOOM 0: ZoomFit to graph the normal distribution. The default values for mean $\mu$ and standard deviation $\sigma$ are 0 and 1.<br><br>The command 2nd DISTR 2:normalcdf(lowerbound, upperbound, $\mu, \sigma$) gives the normal distribution probability that x lies between the lowerbound and the upperbound, when the mean is $\mu$ and the standard deviation is $\sigma$. | Xmin = 29, Xmax = 41, Ymin = 0, Ymax = .2<br><br>```
Plot1 Plot2 Plot3
\Y1■normalpdf(X,
35,2)
\Y2=
\Y3=
\Y4=
\Y5=
\Y6=
```<br><br>```
normalcdf(33,37,
35,2)
       .6826894809
``` |
```

XX. LIMITS

The TI-82, TI-83, and TI-83 Plus calculators are not faultless in evaluating limits, but they are useful in evaluating most limits that we encounter. We can also use them to confirm limits that are evaluated analytically.

LIMITS	EXAMPLE
To find the limit $\lim\limits_{x \to c} f(x)$ for the function f(x):	Evaluate $\lim\limits_{x \to 3} \dfrac{x^2 - 9}{x - 3}$.
1. Enter the function as y_1, and graph the function in a window that contains $x = c$ and the graph of the function where it appears to cross $x = c$.	1. The graph of $y = \dfrac{x^2 - 9}{x - 3}$, with a window containing $x = 3$, follows.
2. Evaluate the function for several values near $x = c$ and on each side of c, by one of the following methods.	
a. TRACE and ZOOM near $x = c$. If the values of y approach the same number L as x approaches c from the left and right, we have evidence that the limit is L.	2. a.
b. Evaluate $y_1(\{c_1, c_2, ...\})$ for values that are very close to, and to the left of, c. This indicates the value of $\lim\limits_{x \to c^-} f(x)$. Repeating this with values very close to, and to the right of, c indicates the value of $\lim\limits_{x \to c^+} f(x)$. If these two limits are the same, say L, then the limit is L.	The y-values appear to approach 6. b.
c. Use TBLSET to start a table near $x = c$ with Δ Tbl very small. If the y-values approach L as the x-values get very close to c from both sides of c, we have evidence that the limit is L.	$\lim\limits_{x \to 3^+} \dfrac{x^2 - 9}{x - 3} = 6$ $\lim\limits_{x \to 3^-} \dfrac{x^2 - 9}{x - 3} = 6$, so $\lim\limits_{x \to 3} \dfrac{x^2 - 9}{x - 3} = 6$ c.
d. Use TBLSET with Indpnt: set to Ask, and enter values very close to, and on both sides of, c. The y-values will approach the same limit as above.	The limit as x approaches 3 appears to be 6. The error at x=3 indicates that f(3) does not exist.

LIMITS WITH PIECEWISE-DEFINED FUNCTIONS	EXAMPLE
To evaluate the limit as $x \to a$ for the piecewise-defined function $y = \begin{cases} f(x) \text{ if } x \le a \\ g(x) \text{ if } x > a \end{cases}$:	Find $\lim_{x \to -5} y$ if $y = \begin{cases} x+7 \text{ if } x \le -5 \\ -x+2 \text{ if } x > -5 \end{cases}$
1. Graph the function, as follows: Under the Y= key, set $y_1 = (f(x))/(x \le a)$ where $\le$ is found under 2nd TEST 6. and set $y_2 = (g(x))/(x > a)$ where > is found under 2nd TEST 3. Use GRAPH or ZOOM to graph the function.	1. Enter $y_1 = (x+7)/(x \le -5)$ and $y_2 = (-x+2)/(x > -5)$
2. Evaluate the function for several values near $x = c$ and on each side of c, by one of the following methods.	2.a.
a. TRACE and ZOOM near $x = c$. If the values of y approach the same number L as x approaches c from the left and right, we have evidence that the limit is L. Use the up or down arrows to TRACE on the correct piece of the function.	The limit from the left does not equal the limit from the right, so the limit does not exist.
b. Use TBLSET to start a table near $x = c$ with Δ Tbl very small. If the y-values approach L as the x-values get very close to c from both sides of c, we have evidence that the limit is L. "ERROR" will occur in the table where the piece y_1 or y_2 does not exist.	b. $\lim_{x \to -5^-} y = 2$, $\lim_{x \to -5^+} y = 7$, so $\lim_{x \to -5} y = DNE$. Note that $y(-5) = 2$.
c. Use TBLSET with Indpnt: set to Ask, and enter values very close to, and on both sides of, c. The y-values will approach the same limit as above.	c.

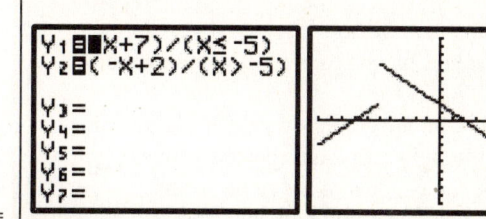

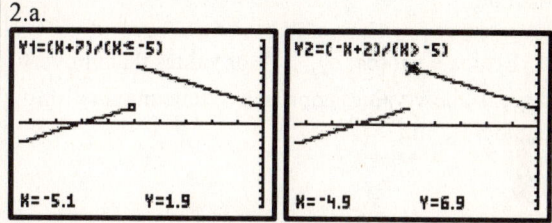

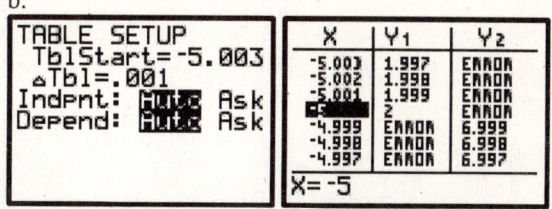

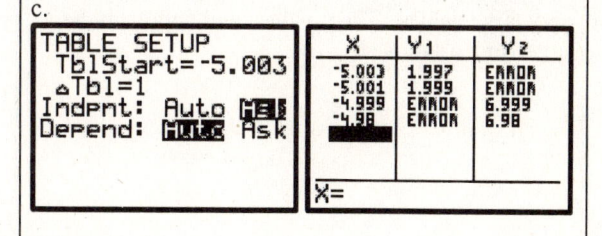

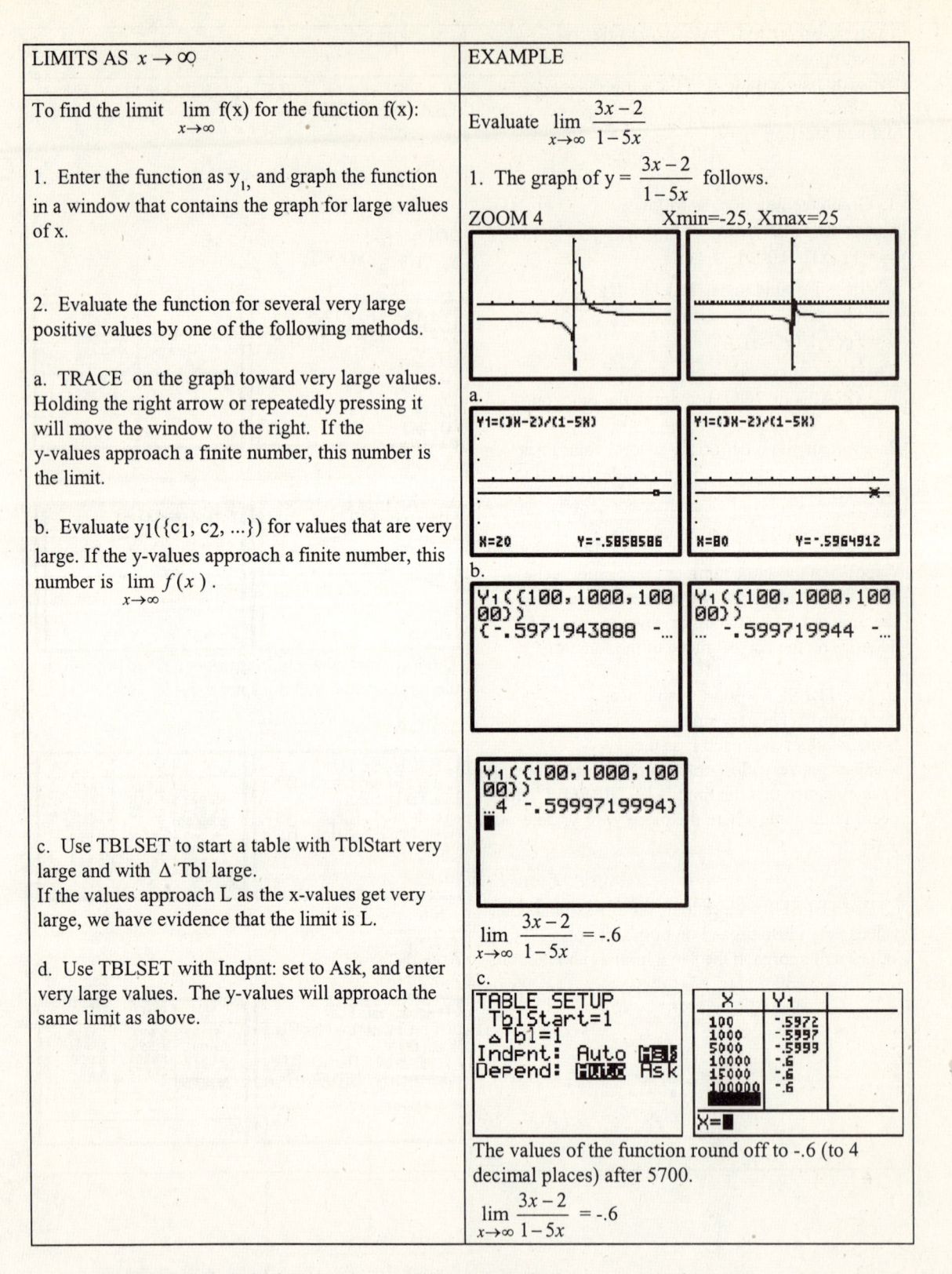

LIMITS AS $x \to \infty$	EXAMPLE
To find the limit $\lim\limits_{x \to \infty} f(x)$ for the function f(x):	Evaluate $\lim\limits_{x \to \infty} \dfrac{3x-2}{1-5x}$
1. Enter the function as y_1, and graph the function in a window that contains the graph for large values of x.	1. The graph of $y = \dfrac{3x-2}{1-5x}$ follows. ZOOM 4 Xmin=-25, Xmax=25
2. Evaluate the function for several very large positive values by one of the following methods.	
a. TRACE on the graph toward very large values. Holding the right arrow or repeatedly pressing it will move the window to the right. If the y-values approach a finite number, this number is the limit.	a.
b. Evaluate $y_1(\{c_1, c_2, ...\})$ for values that are very large. If the y-values approach a finite number, this number is $\lim\limits_{x \to \infty} f(x)$.	b.
c. Use TBLSET to start a table with TblStart very large and with Δ Tbl large. If the values approach L as the x-values get very large, we have evidence that the limit is L.	$\lim\limits_{x \to \infty} \dfrac{3x-2}{1-5x} = -.6$
d. Use TBLSET with Indpnt: set to Ask, and enter very large values. The y-values will approach the same limit as above.	c. The values of the function round off to -.6 (to 4 decimal places) after 5700. $\lim\limits_{x \to \infty} \dfrac{3x-2}{1-5x} = -.6$

In a similar fashion, the calculator can be used to investigate limits as x approaches $-\infty$.

XXI. NUMERICAL DERIVATIVES

There are two ways to find the derivative of a function at a specified value of x. One uses the graph of the function and one uses the numerical derivative operation of the MATH menu. Both of these methods follow.

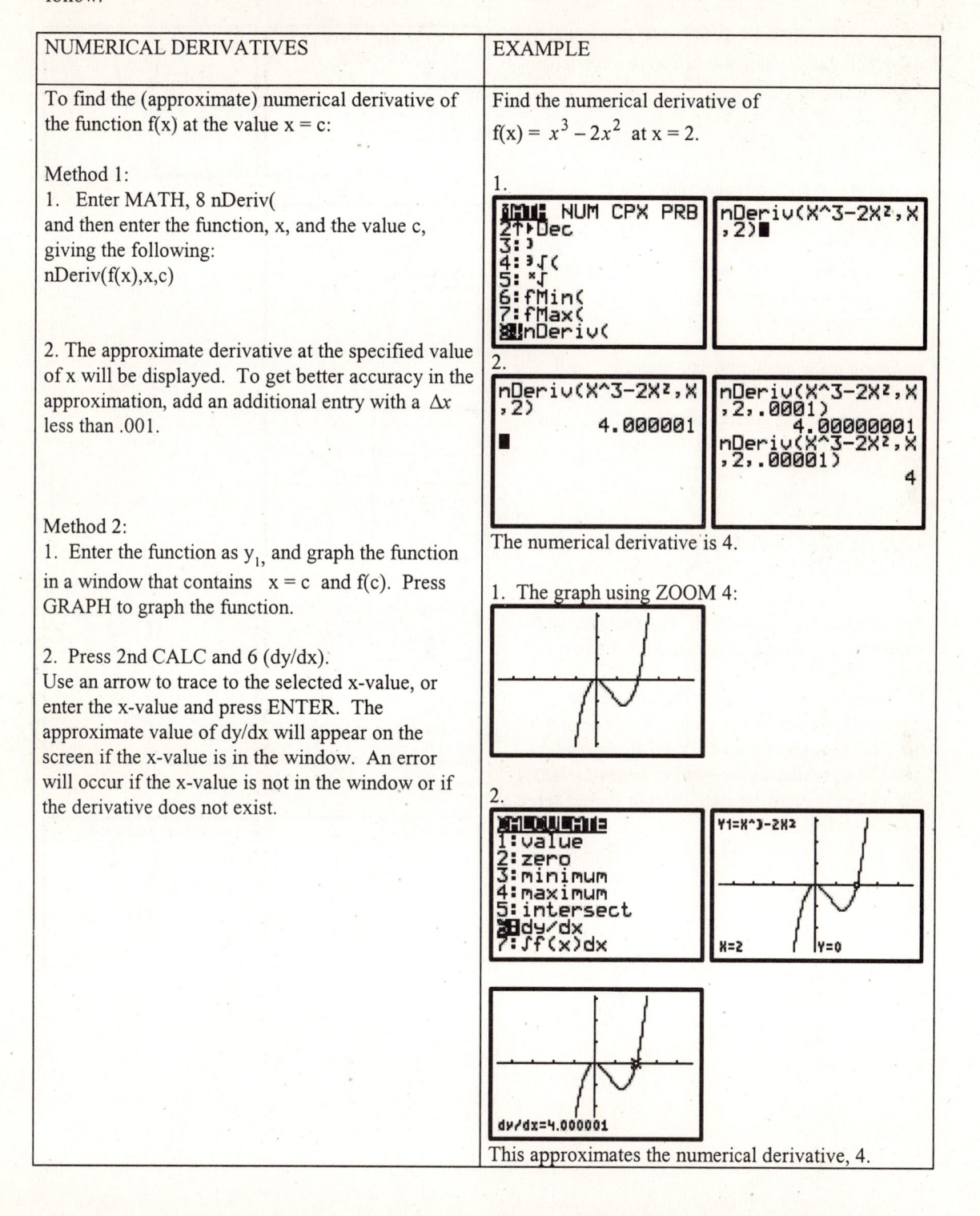

NUMERICAL DERIVATIVES	EXAMPLE
To find the (approximate) numerical derivative of the function f(x) at the value x = c: Method 1: 1. Enter MATH, 8 nDeriv(and then enter the function, x, and the value c, giving the following: nDeriv(f(x),x,c) 2. The approximate derivative at the specified value of x will be displayed. To get better accuracy in the approximation, add an additional entry with a Δx less than .001. Method 2: 1. Enter the function as y_1, and graph the function in a window that contains x = c and f(c). Press GRAPH to graph the function. 2. Press 2nd CALC and 6 (dy/dx). Use an arrow to trace to the selected x-value, or enter the x-value and press ENTER. The approximate value of dy/dx will appear on the screen if the x-value is in the window. An error will occur if the x-value is not in the window or if the derivative does not exist.	Find the numerical derivative of $f(x) = x^3 - 2x^2$ at x = 2. 1. 2. The numerical derivative is 4. 1. The graph using ZOOM 4: 2. This approximates the numerical derivative, 4.

After you have found the derivative of a function, you can use its graph and the graph of the numerical derivative of the function to check your work.

CHECKING A DERIVATIVE	EXAMPLE
To check the correctness of a derivative $f'(x)$ of the function f(x):	Verify that the derivative of $f(x) = x^3 - 2x^2$ is $f'(x) = 3x^2 - 4x$

1. Enter the derivative $f'(x)$ that you found as y_1, and graph this derivative function in a convenient window.

1.

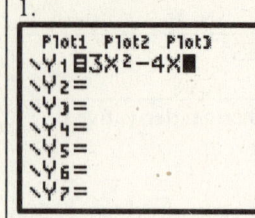

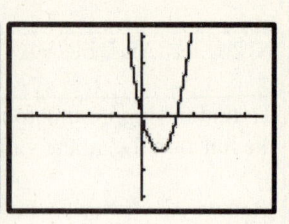

2. Press Y = and enter the following in y_2: nDeriv(f(x),x,x).

2.

3. Graph using an appropriate window. Both graphs will appear. If the second graph lies on top of the first, then the derivatives agree, and your solution checks.

3.

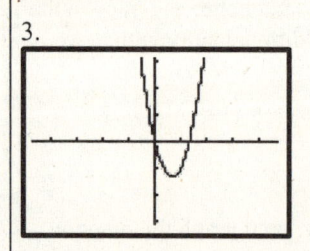

4. On the TI-83 or TI-83 Plus, move the cursor to the left of $y_2 =$ and press ENTER to get a thicker \, which indicates the graph will be drawn thicker than normal. Press GRAPH. The second graph will now be thicker as it graphs over the first.

4.

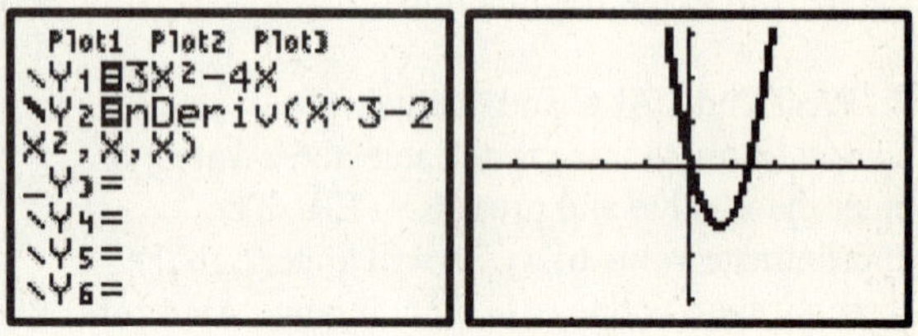

5. To further verify that the derivatives agree (especially on the TI-82), move from one graph to the other by using the up or down arrow, and use TRACE to evaluate them both at several x-values. The values may not be identical, but should agree when rounded.

5.

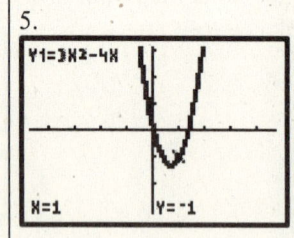

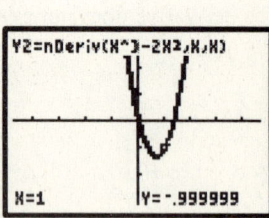

FINDING AND TESTING SECOND DERIVATIVES	EXAMPLES
To find the second derivative of a function at a given value of x at x = c: 1. Enter the function as y_1. Enter nDeriv(y_1,x,x) as y_2. 2. The second derivative of the function given by Y_1 is approximated at x = c with nDeriv(Y_2,x,c) To check the correctness of a derivative $f''(x)$ of the function f(x): 1. Enter the function as y_1. Enter nDeriv(y_1,x,x) as y_2. 2. Enter nDeriv(y_2,x,x) as y_3. 3. Enter the second derivative $f''(x)$ that you found as y_4. (On the TI-83, move the cursor to the left of Y4 = and press ENTER to get a thicker \, which indicates the graph will be drawn thicker than normal.) 4. Turn off the equations for y_1 and y_2. Graph y_3 and y_4 in the same window. If the second graph lies on top of the first, then the derivatives agree, and your solution checks.	Find the second derivative of $f(x) = x^3 - 2x^2$ at x = 2. 1. 2. Thus $f''(2) = 8$. 1. 2. 3. 4.

81

XXII. CRITICAL VALUES

The values of x that make the derivative of a function 0 or undefined are critical values of the function. We can find where the derivative is 0 by finding the zeros of the derivative function (that is, the x-intercepts of the derivative function).

CRITICAL VALUES	EXAMPLE
To find or approximate values of x that make the derivative of f(x) equal to 0: I. Find the derivative of f(x). II. Find the values of x that make the derivative 0: Method 1 1. Enter the equation of the derivative as y_1. 2. Find where $y_1 = 0$ by one of the following: a. Finding the x-intercepts of y_1 by using TRACE. b. Finding the zeros of y_1 by using 2nd CALC, 2 :zero (root on the TI-82). c. Use 2nd TBLSET and 2nd TABLE to find values of x that give y = 0. Method 2 (On the TI-83 or TI-83 Plus) 1. Press MATH, 0: Solver 2. Press ENTER and the up arrow, and then enter the equation of the derivative on the right of 0=. 3. Press the down arrow to put the cursor on the variable, and press ALPHA, SOLVE, ENTER. A solution will appear if one exists. Changing values set equal to the variable and using ALPHA, SOLVE, ENTER will give other solutions if they exist.	Find the values that make the derivative of $f(x) = \dfrac{x^3}{3} - 4x$ equal to 0. I. The derivative is $f'(x) = x^2 - 4$. 1. 2. a. TRACE gives one x-intercept to be 2. b. Thus 2nd CALC Zero gives the zero -2. c. 1. 2. 3. The solutions (zeros) are 2 and -2.

XXIII. RELATIVE MAXIMA AND RELATIVE MINIMA

RELATIVE MAXIMA AND RELATIVE MINIMA USING THE DERIVATIVE	EXAMPLE
To find the relative maximum and or relative minimum of a polynomial function: 1. Find the derivative of the function, and enter the equation of the derivative as y_2. 2.a. Graph y_2 and find the values of x that make the derivative 0 or undefined, using TRACE, ZERO, or TABLE. (No value of x will make the derivative of a polynomial undefined.) 3. Enter the function as y_1. 4. Use TBLSET and TABLE to find derivatives near, and to the left and right of the zeros. of the derivative: a. If the derivative, y_2 is positive to the left of x = c, 0 at c and negative to the right of c, the function has a relative maximum at x = c. The y-value, y_1, where x = c is the relative maximum that occurs there. b. If the derivative, y_2 is negative to the left of x = c, 0 at c and positive to the right of c, the function has a relative minimum at x = c. The y-value, y_1, where x = c is the relative minimum that occurs there. c. If the derivative does not have opposite signs on opposite sides of x = c, there is no maximum nor minimum, and a horizontal point of inflection occurs. 5. Graph the function to confirm the relative maximum and relative minimum occur where you have found them.	Find the relative maximum of $$f(x) = \frac{x^3}{3} - 4x$$ 1. $f'(x) = x^2 - 4$ 2. TBLSET -3, Δ TBL 1 The derivative in column y_2 is 0 at x = -2 and x = 2. 4. Using TABLE with the derivative in column Y_2 gives a. The values of y_2 change from positive to 0 at x = -2 to negative, so a relative maximum occurs at x = -2; and from y_1, y = 5.3333 when x = -2, so a relative maximum is at (-2, 16/3). b. y_2 changes from negative to 0 at x = 2 to positive, so a relative minimum occurs at x = 2; and from y_1, y = -5.3333 when x = 2, so a relative minimum is at (2, -16/3). 5. max @ (-2, 16/3), min @ (2, -16/3)

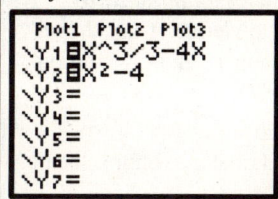

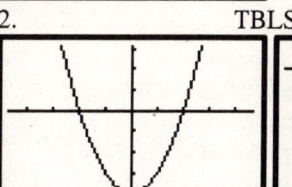

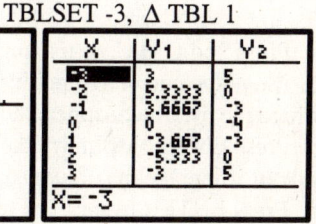

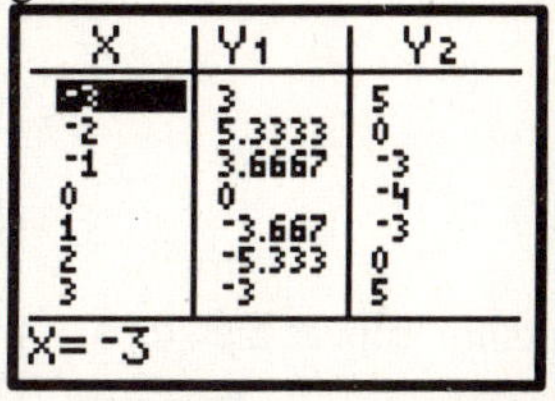

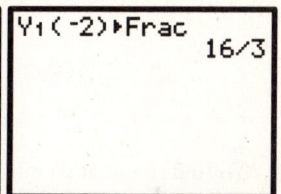

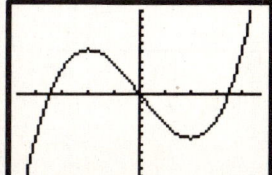

RELATIVE MAXIMA AND RELATIVE MINIMA USING MAXIMUM OR MINIMUM	EXAMPLE
I. To find the relative maximum of a polynomial function: 1. Enter the equation of the function as y_1. 2. Select a window that includes all possible "turns" in the graph, using knowledge of the shapes of polynomial functions. Graph y_1, and use ZOOM, Zoom Out to find all "turns." 3. If there appears to be a relative maximum, locate it as follows: a. Press 2nd CALC, 4 (maximum). b. Move the cursor to a point on the left side of where the maximum appears to occur. c. Press ENTER and move the cursor to the right side of where the maximum appears to occur. d. Press ENTER twice. The resulting point is an approximation of the observed relative maximum. II. To find the relative minimum, repeat the steps above using 2nd CALC, 3 (minimum).	Find the relative maximum of $f(x) = \dfrac{x^3}{3} - 4x$ 1. 2. 3. a. b. c. d. The relative maximum, found with calculus, is really $y = 16/3$ at $x = -2$. II. The relative minimum is $y = -16/3$ at $x = 2$.

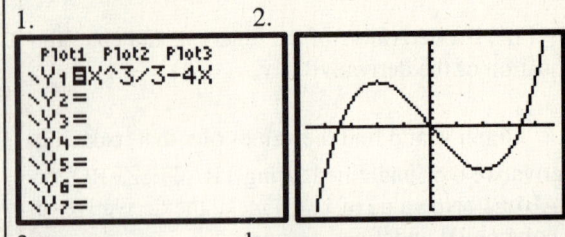

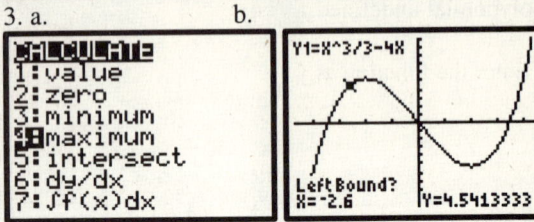

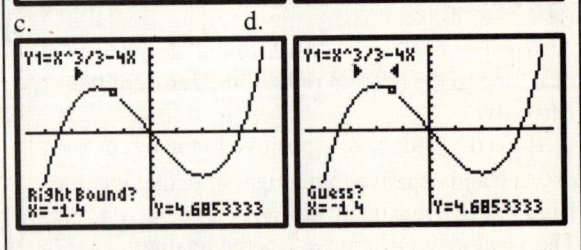

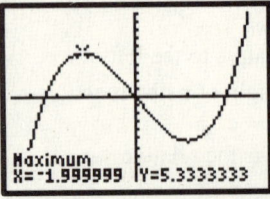

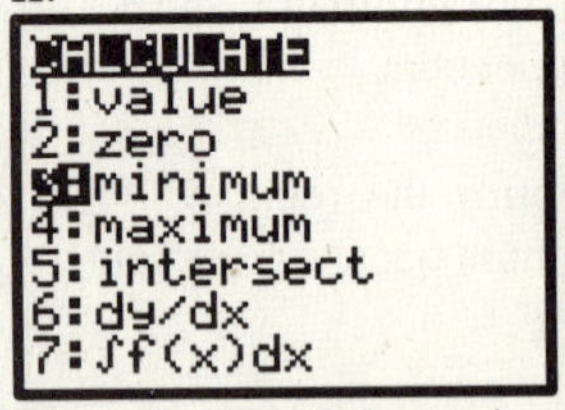

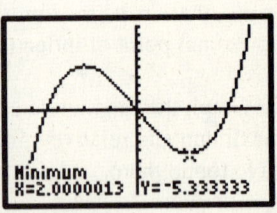

84

UNDEFINED DERIVATIVES AND RELATIVE EXTREMA	EXAMPLE
If the derivative of f(x) is undefined at x = c, and if f(c) exists, then find a relative maximum or minimum as follows: 1. Enter the equation of the derivative as y_2. 2. Graph y_2 and find the values of x that make the derivative undefined. (Use TRACE, ZERO, or TABLE.) 3. Enter the equation of the function as y_1. 4. If y_1 exists where the derivative is undefined, use TBLSET and TABLE to find derivatives near, and to the left and right of this x-value. a. If the derivative, y_2 is positive to the left of x = c, undefined at c and negative to the right of c, the function has a relative maximum at x = c. The y-value, y_1, where x = c is the relative maximum that occurs there. b. If the derivative, y_2 is negative to the left of x = c, undefined at c and positive to the right of c, the function has a relative minimum at x = c. The y-value, y_1, where x = c is the relative minimum that occurs there. c. If the derivative does not have opposite signs on opposite sides of x = c, there is no maximum nor minimum, and a vertical point of inflection occurs. 5. Graph the function to confirm the relative maximum and relative minimum occur where you have found them.	Find the relative maximum and or relative minimum of $f(x) = (x-1)^{2/3} + 2$. 1. 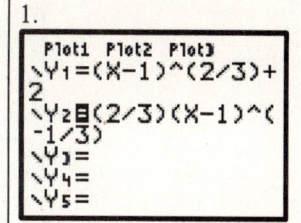 2. 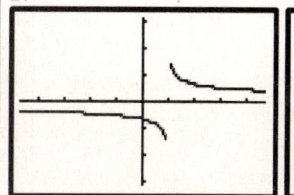 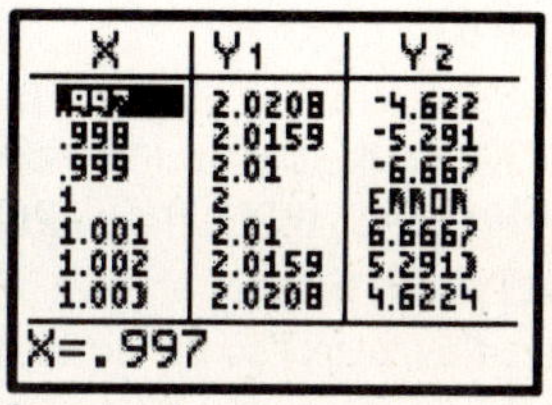 The derivative is undefined at x = 1. 4. The values of y_2 change from negative to undefined at x = 1 to positive, so a relative minimum occurs at x = 1. Using the table that includes y_1 shows that the relative minimum is y = 2 where x = 1. 5. The graph of the function is rel min at (1,2).

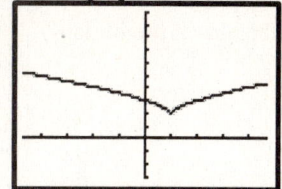

XXIV. INDEFINITE INTEGRALS

CHECKING INDEFINITE INTEGRALS	EXAMPLES
To check an indefinite integral with fnInt:	Find the integral of $f(x) = x^2$ and check the result with fnInt.
1. Enter the integral of f(x) (without the +C) as y_1 under the Y= menu.	1. The integral is $\dfrac{x^3}{3} + C$. Enter $y_1 = x\char94 3/3$ under the Y= menu.
2. Move the cursor to y_2, press MATH, 9: fnInt(. Enter f(x),x,0,x) so the equation is $y_2 = $ fnInt(f(x),x,0,x).	2.
3. Press GRAPH with an appropriate window. If the second graph lies on top of the first, the graphs agree and the computed integral checks.	3. The second graph lies on top of the first, which indicates that the integrals agree.
4. On the TI-83, pressing ENTER with the cursor to the left of y_2 changes the thickness of the graph of the second graph, making the fact that it lies on top of the first more evident.	4.

FAMILIES OF FUNCTIONS SOLVING INITIAL VALUE PROBLEMS	EXAMPLES
The indefinite integral of the function f(x) has the form F(x) + C, where the derivative of F(x) is f(x). Thus the indefinite integral gives a family of functions, one for each value of C. Different values of C give different functions. To graph some of them: 1. Integrate f(x). 2. Enter equations in the equation editor, using different values for C. 3. Press GRAPH with an appropriate window. The graphs will be graphs of y = F(x) shifted up or down, depending on C. If a value of x and a corresponding value of y are given for the integral of a function, This "initial value" can be used to solve for C and thus to find the one function that satisfies the conditions. To find this function: 1. Integrate the function f(x). 2. Press MATH, 0:Solver and press the up arrow to see EQUATION SOLVER. 3. Set 0 equal the integral minus y, getting 0 = F(x) + C - y, and press the down arrow. 4. Enter the given values of x and y, place the cursor on C, and press ALPHA, SOLVE (ENTER). The value of C will appear. Replace C with this value to find the function satisfying the conditions.	Find the integral of $f(x) = 2x - 4$ and graph the integrals for $C = 0$, $C = 1$, $C = -2$, and $C = 3$. The integral of $f(x) = 2x - 4$ is $f(x) = x^2 - 4x + C$. The equation editor showing these equations and the graphs of these equations are shown below. 2. 3. 1. The integral of $f(x) = 2x - 4$ is $y = x^2 - 4x + C$. 2. 3. 4. The unique function is $y = x^2 - 4x + 3$. Its graph is shown above.

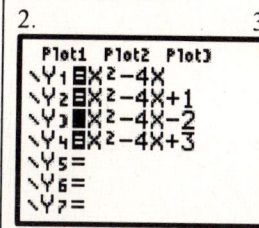

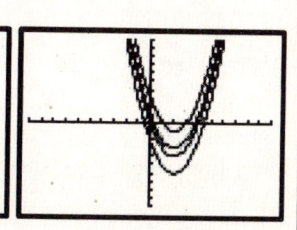

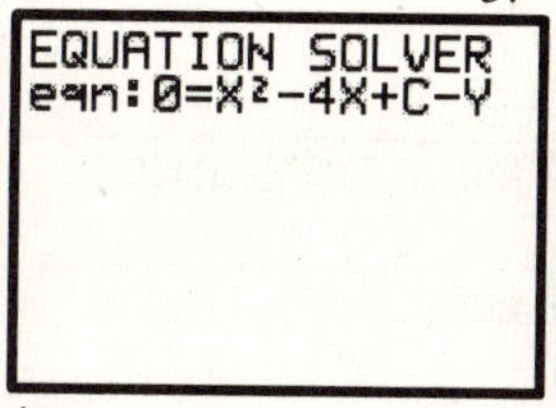

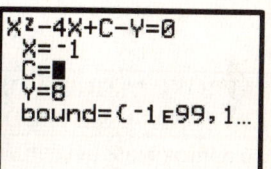

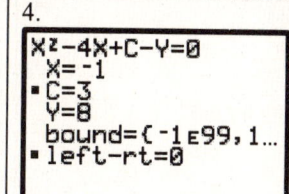

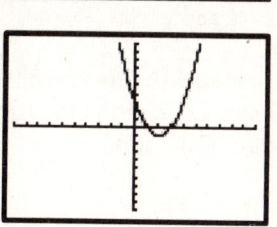

XXV. DEFINITE INTEGRALS

APPROXIMATING A DEFINITE INTEGRAL - AREAS UNDER CURVES	EXAMPLES
To find the area under the graph of y = f(x) and above the x-axis: 1. Enter f(x) under the Y= menu, and press GRAPH with an appropriate window. 2. Press 2nd CALC and 7: $\int f(x)dx$. 3. Press ENTER. Move the cusor to, or enter, the lower limit (the left x-value). 4. Press ENTER. Move the cursor to, or enter the upper limit (the right x-value). 5. Press ENTER. The area will be displayed.	EXAMPLE Find the area under the graph of $f(x) = x^2$ from x = 0 to x = 3.
APPROXIMATING A DEFINITE INTEGRAL - ALTERNATE METHOD	EXAMPLE
To approximate the definite integral of y = f(x) in the interval between x = a and x = b: 1. Press MATH, 9: fnInt(. Enter f(x),x,a,b) so the display shows fnInt(f(x),x,a,b). 2. Press ENTER to find the approximation of the integral. 3. The approximation may be made closer than that in step 3 by adding a fifth argument with a number (tolerance) smaller than 0.00001.	Approximate the definite integral of $f(x) = 4x^2 - 2x$ from x = -1 to x = 3. This approximation is not improved. The exact integral is 88/3 = 29 1/3.

AREA BETWEEN TWO CURVES	EXAMPLES
To find the area enclosed by the graphs of two functions: 1. Enter one equation as y_1 and the second as y_2. Press GRAPH using an appropriate window. 2. Find the x-coordinates of the points of intersection of the graphs. Use 2nd CALC 5: intersect. 3. Determine visually which graph is above the other over the interval between the points of intersection. 4. Press MATH, 9: fnInt(. Enter f(x),x,a,b so the display shows fnInt(f(x),x,a,b) where f(x) is $y_2 - y_1$ if the graph of y_2 is above the graph of y_1 between a and b, or $y_1 - y_2$ if y_1 is above y_2. The area between the graphs can also be found using 2nd CALC, $\int f(x)dx$. 1. Enter $y_3 = y_2 - y_1$ where y_2 is above y_1. 2. Turn off the graphs of y_1 and y_2 and graph y_3 with a window showing where $y_3 > 0$. 3. Press 2nd CALC and 7: $\int f(x)dx$. 4. Press ENTER and select the lower limit (the left x-intercept). 5. Press ENTER. Move the cursor to, or enter, the upper limit (the right x-intercept). 6. Press ENTER. The area will be displayed.	Find the area enclosed by the graphs of $y = 4x^2$ and $y = 8x$. 1. 2. Graphs intersect at x = 0 and x = 2. 3. $y = 8x$ is above $y = 4x^2$ in the interval from 0 to 2. 4. fnInt(Y2-Y1,X,0,2) 5.333333333 1. 2. 3. 4. 5. 6.

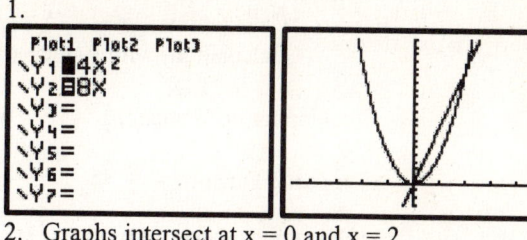

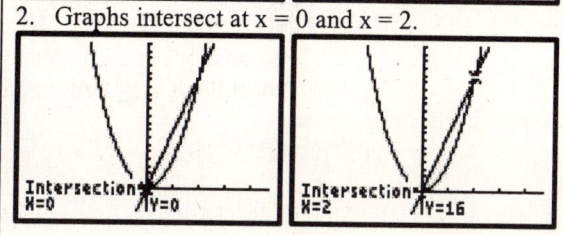

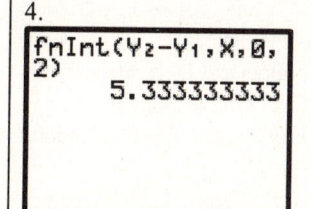

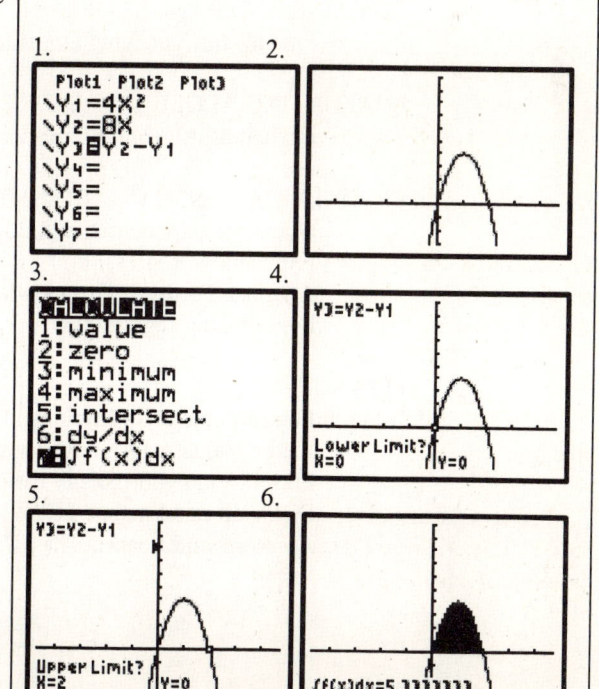

INDEX

AREA	89
Area Between Two Curves	89
CALCULATIONS WITH THE TI-83, TI-83 PLUS, AND TI-82	14
Calculations	14
Calculations with Radicals and Rational Exponents	15
Combinations of Functions	26
Composite Functions	27
COUNTING AND PROBABILITY	70
Permutations and Combinations	70
Probability Using Permutations and Combinations	71
CRITICAL VALUES	82
DEFINITE INTEGRALS	88
Approximating a Definite Integral-Areas Under Curves	88
Approximating a Definite Integral-Alternate Method	88
Area Between Two Curves	89
DERIVATIVES	79
Numerical Derivatives	79
Checking a Derivative	80
Finding and Testing Second Derivatives	81
Determinants	47
DOMAINS AND RANGES OF FUNCTIONS	25
Finding and Verifying Domains and Ranges of Functions	25
EVALUATING ALGEBRAIC EXPRESSIONS	16
Evaluating Algebraic Expressions Containing One or More Variables	16
EXPONENTIAL AND LOGARITHMIC FUNCTIONS	57
Graphs of Exponential and Logarithmic Functions	57
Exponential Regression	59
Alternate Forms of Exponential Functions	60
Logarithmic Regression	61
FINANCE	65
Future Value of an Investment	65
Future Values of Annuities and Payments Into Sinking Funds	66
Present Value Formulas - Evaluating with TABLE	67
SOLVER and Finance Formulas on the TI-83 and TI-83 Plus	68
Annuities and Loans - The FINANCE Application on the TI-83 and TI-83 Plus	69

FUNCTIONS 22
    Evaluating Functions with TRACE, Value 22
    Evaluating Functions with TABLE 23
    Evaluating Functions with y-vars 24
    Inverse Functions 58
    Piecewise-Defined Functions 40

GRAPHING EQUATIONS AND FUNCTIONS 17
    Using a Graphing Calculator to Graph an Equation 17
    Viewing Windows 18
    Finding y-values for Specific Values of x 19
    Graphing Equations on Paper 20
    Using a Graphing Calculator to Graph Equations Containing y^2 21
    Graphing Piecewise-Defined Functions 40
    Graphs of Exponential and Logarithmic Functions 57
    Graphs of Special Functions 37

INTEGRALS
    Checking Indefinite Integrals 86
    Approximating a Definite Integral-Areas Under Curves 86
    Approximating a Definite Integral-Alternate Method 88
    Area Between Two Curves 88
    Families of Functions 89

INTERCEPTS OF GRAPHS 28
    Finding or Approximating y- and x-Intercepts of a Graph Using Trace 28
    Using CALC, ZERO to Find the x-intercepts of a Graph 29

INVERSE FUNCTIONS 58

LIMITS 76
    Limits 76
    Limits with Piecewise-Defined Functions 77
    Limits as $X \to \infty$ 78

LINEAR PROGRAMMING 56
    Graphical Solution of Linear Programming Problems 56
    Markov Chains 72

MATRICES 43
    Entering Data into Matrices 43
    Determinant of a Matrix 47
    Identity Matrix 43
    Inverse Matrices 46
    Operations with Matrices 44
    Multiplying Two Matrices 45
    Solving Systems of Linear Equations with Unique Solutions 48
    Solution of Systems of Three Equations in Three Variables 49
    Solution of Systems - Reduced Echelon Form on the TI-83 and TI-83 Plus 50
    Solution of Systems - Non-Unique Solutions 51
    Transposes of Matrices 47

MAXIMA AND MINIMA
    See Relative Maxima and Minima

MODELING DATA 42
Modeling Data 42

NUMERICAL DERIVATIVES 79
Numerical Derivatives 79
Checking a Derivative 80
Finding and Testing Second Derivatives 81

PERMUTATIONS AND COMBINATIONS 70
Permutations and Combinations 70
Probability Using Permutations and Combinations 71

PIECE-WISE DEFINED FUNCTIONS 40
Graphing Piecewise-Defined Functions 40

POLYNOMIAL FUNCTIONS
Relative Maxima And Minima Using Maximum/Minimum 84

PROBABILITY
Probability Using Permutations and Combinations 71

QUADRATIC FUNCTIONS 38
Approximating the Vertex of a Parabola with TRACE 38
Finding the Vertex of a Parabola with CALC, MAXIMUM or MINIMUM 39
Finding y- and x-Intercepts of a Graph Using TRACE 28
Using CALC, ZERO to Find the x-intercepts of a Graph 29

RADICALS 15
Calculations with Radicals and Rational Exponents 15

RELATIVE MAXIMA AND RELATIVE MINIMA 83
Relative Maxima and Minima Using The Derivative 83
Relative Maxima And Minima Using Maximum/Minimum 84
Undefined Derivatives and Relative Extrema 85

SCATTERPLOTS 41
Scatterplots of Data 41

SEQUENCES 62
Evaluating a Sequence 62
Arithmetic sequences - nth Terms and Sums 63
Geometric Sequences - nth Terms And Sums 64

SOLVING EQUATIONS 30
Using TRACE to Find or Check Solutions of Equations 30
Solving Equations with the ZERO (or ROOT) Method 31
Solving Equations Using the Intersect Method 32
Solving Equations with SOLVER on the TI-83 and TI-83 Plus 33
Solving an Equation for One of Several Values with SOLVER 33

SOLVING INEQUALITIES ... 52
    Solving Linear Inequalities ... 52
    Solving Systems of Linear Inequalities ... 53
    Solving Quadratic Inequalities ... 54
    Solving Quadratic Inequalities - Alternate Method ... 55

SOLVING SYSTEMS OF EQUATIONS ... 34
    Points of Intersection of Graphs - Solving a System of Two Linear Equations Graphically ... 34
    Solution of Systems of Equations Using the Intersect Method ... 35
    Solution of Systems of Equations – Finding or Approximating Using TABLE ... 36
    Solution of Systems of Equations: Non-Unique Solutions ... 51
    Solving Systems of Linear Equations with Unique Solutions ... 48
    Solution of Systems of Three Equations in Three Variables ... 49
    Solution of Systems - Reduced Echelon Form on the TI-83 and TI-83 Plus ... 50

STATISTICS ... 73
    Descriptive Statistics ... 73
    Histograms ... 74
    Probability Distributions ... 75

VERTICES OF GRAPHS OF QUADRATIC FUNCTIONS ... 38
    Approximating the Vertex of a Parabola with TRACE ... 38
    Finding the Vertex of a Parabola with CALC, MAXIMUM or MINIMUM ... 39